HOW TO
INNOVATE
— ON —
PURPOSE

CHASE FEWER IDEAS.
BUILD BETTER PRODUCTS.

First Edition

DIANE PIERSON

innovateonpurpose.com

For information, please email Diane Pierson at dpierson@innovateonpurpose.com.

eBook ISBN: 979-8-9886601-0-1

Print Book ISBN: 979-8-9886601-1-8

CONTENTS

FIGURES

Chapter 1
INTRODUCTION

THAT FAMOUS MARKET-DRIVEN INNOVATOR

A book on how to innovate should start with a great example of market-driven innovation, and I have one for you. I don't even have to name the company! You can all think of who I'm talking about when I say it is "a massive company that came out of nowhere, using technology and innovative business practices to dominate the retailing landscape without even having an actual store!" right? The company that disrupted the established way of selling consumer products, of *buying* consumer products; the company that leveraged emerging technology and exploited the weaknesses of the current competitors to win; the company that expanded far beyond its original line of business to be synonymous with "shopping."

I am, of course, talking about Sears.

Sears, Roebuck and Co. Some of you are asking, "Who?" Some of you know who I'm talking about but only remember Sears as a dowdy store at the dead-end mall you stopped going to years ago. But in the first half of the twentieth century, Sears was a disruptive, cutting-edge innovator. Using the power of a catalogue, railroads, and the US Postal Service, Sears was ubiquitous for decades *without having a physical location*—50 years before Amazon. Further, Sears reinvented the workflow of shopping for, buying, and receiving goods, *without selling anything new.*

They were laser-focused on a market: Americans in rural areas beholden to the limited stock and subjective credit terms of a local general store. And yes, you could buy nearly anything at Sears—from a dress to a drill to a darn fine house. Much more than your not-so-local general store could ever stock. And you didn't have to hitch up the horse and buggy (assuming you had these luxuries) to lug them home. You ordered what you wanted and had it delivered through that amazing innovation: US Mail Rural Home Delivery.

As those rural customers started to buy automobiles, Sears flipped their own model of success and opened what were essentially reimagined general stores that disrupted their own catalogue business. The catalogue continued to be mailed, but its purpose morphed from order form to marketing event. The Sears "Wishbook," as it was informally called, delivered to your mailbox shortly after Thanksgiving and used as a how-to manual by children from the 1930s to the 1990s to create their Christmas present lists for Santa Claus.

THE FALL OF AN INNOVATOR

But time moved on, and Sears did not. As of November 2022, Sears had emerged from Chapter 11 bankruptcy but was liquidating remaining assets. At this writing, there are only 20 of the former 3,500 Sears stores remaining, and by the time this book is available there will likely be even fewer. Ask a child today whether they used the Wishbook to make their list to Santa, and you'll get a blank stare—possibly from their parents as well!

There are myriad reasons for the fall of Sears, but certainly failure to change with a changing market—let alone lead the way into retailing's future—is a fundamental contributing factor.

The rise of Sears is legendary—and the fall, a cautionary tale. Both reinforce the need for companies to master market-driven innovation. To do that, you need to intentionally identify a market and work to understand it so you can deliver products they need and want.

And it's not all about the product. Sears sold the same things the general stores did when they disrupted the general stores with such breathtaking success. Best Buy, Walmart, and of course Amazon sold the exact same things Sears did to disintermediate Sears. Frequently it's not about building a better mousetrap; it's about positioning, pricing, selling, and delivering that mousetrap in the way your market wants it. It's about synthesizing trends, technology, and tools perhaps currently unused or unexpressed. These facets of "product" differentiation and growth are often overlooked, for the scary-simple reason that there's no one *responsible* for looking at it from that perspective. More on that later.

But it *is* all about the market. For any organization, your first questions should be, "Which markets do we serve?" and "What do we want to be to them?" Before you can be an innovator, you must know who you're innovating *for*. You need to have a clear focus on who you're trying to serve and a deep understanding of that group without regard to any given product you're trying to offer. Only then can you drill down into discovery of problems to solve. Market-driven innovation isn't reactive, or a contradiction in terms, when you see it from this angle. In fact, using this lens, you can see how that most sought-after innovation claim—the "creating of a market"—is the deepest level of market understanding.

Because the needs of the world and the tools available to fulfill them are constantly changing, the mandate for innovation is also constant. A product that was hailed as genius a decade ago may be unusable now—something RIM found out about their Blackberry smartphone the hard way. The high-priority need of a market today could be forgotten tomorrow. How much hand sanitizer have you bought in 2023 versus 2020? The only constant is that, at any given point in time, there are groups of people with similar problems who want them solved. Your challenge is to establish a workflow that makes sure you're staying on top of the changes and responding to them.

Innovation is a journey. Too many organizations are wandering through that journey without a map—without a plan. Like most journeys conducted in this way, finding what you and your markets are looking for ends up being pure luck—and too often results in a lot of time and effort wasted, frustrating ourselves and our potential customers.

WHY DOES INNOVATION FAIL?

Every organization is trying to innovate, but most are going about it in ways more likely to fail than succeed. And not just fail but spend enormous amounts of money, time, and customer goodwill to do so.

Why don't most companies innovate well? There's so much market data available! Here are some statistics from a March 2021 article in *Tech Jury*:

- 97.2% of organizations are investing in big data and AI.
- Poor data quality costs the US economy up to $3.1 trillion yearly.
- 95% of businesses cite the need to manage unstructured data as a problem.

Source: https://techjury.net/blog/big-data-statistics/#gref.

With so much information being collected—why is there so little understanding? So many off-point marketing communications? So many products no one buys? Clearly, it's not that nobody's listening to their markets but rather that they're listening in incomplete or unactionable ways and not acting on what they do learn. Maybe data isn't aggregated, so no one has a complete picture of the market. Maybe the knowledge gleaned isn't communicated to the right people. But very often, it's simply not leveraged in the innovation workflow.

There's another, larger issue at play. In my experience, product teams don't consider innovation opportunities to be something they *can* find in the market. They conflate innovation with inspiration and good luck, or something only a few geniuses can aspire to. It's a thunderbolt, a brainwave. Almost … magic. Certainly not something you can go *looking* for.

The legends around innovation are largely apocryphal, which makes them dangerous. These legends make product teams believe that innovation comes from their imaginations, instead of the market. The idea that random luck or the isolated brainwave drives innovation leads product teams to reject systematic, disciplined cycles of market research. Instead, they waste a lot of money and time building "innovative" product guesses that no one buys.

Legends are fun, but innovating on purpose is better. Scratch the surface of these legends and their leaders—Apple's Steve Jobs, Reed Hastings of Netflix—and you'll see that there were signals from the market that led to their success. Cultural trends, emerging technology, and sometimes outright demands for change that these visionaries took advantage of. These signals can be identified by any organization, in any market, if you listen in a way designed to find them. You may be doing market research, but you're probably not going to find the Next Big Thing the way you're doing it today.

As a matter of fact—is it really the Next Big Thing that your leadership is looking for? If we limit the definition of "innovation" to a product, wholly new, that nobody even thought to ask for, we're dismissing a lot of potential solution improvement and market opportunities. These out-of-nowhere genius products may not even be what your leadership wants you to focus on.

Innovation isn't about the Big Idea, it's about the **Right** Idea. Innovation fails when your company is focused on expanding relationships with existing customers, through expansion of existing products, and you're swinging for the fences with fabulous but off-point ideas to chase brand-new markets with the next new thing. Without alignment on what type of innovation we're aspiring to as a company, we can spin our wheels.

Several years ago, I came to an organization that had a robust ideation process in place. Once a year, anyone in the organization could pitch a new product idea to the executive team. The execs (of which I was one), would review all ideas, and finalists would present an expanded version of their idea to the execs, who would choose the ideas to go after.

Nearly everyone in the company teamed up to pitch ideas, spending hours of their evenings and weekends putting together their thoughts. Some of these ideas were modest, point-release-type updates to existing products, and some were highly futuristic, risky new directions that could make the company millions by opening new markets. The executive team received and read over 100 of these pitches and held hours-long meetings to review and choose finalists. Sounds exciting! It was a disaster.

One problem was that we had resources for about three new projects every year. Another was that our CEO had a very clear idea of what those projects should look like to drive the strategy of the organization, but the team didn't get (or, to be fair, ask for) that detail. At the end of the day, the team was burning itself out pitching ideas that were **never going to be considered** because they didn't align with where the company was trying to go.

You **can** grow with or even lead your markets over time, open new markets, and build amazing and successful new products, but you need (1) a plan and (2) focus to drive efficiency while keeping us free to find meaningful opportunities to drive the stated goals. And everyone in the organization needs to understand how those strategic objectives apply to the job they do every day.

"Innovation by brainwave" is fun to talk about, but it's risky and expensive—not to mention largely mythical. To *innovate on purpose*, you must look for *inspiration* on purpose. That means creating a plan to find it.

Chances are, your company spends a fortune on focus groups, user conferences, and surveys, all to gather market data. But are you aggregating that data to find patterns? Are you using that data when it comes time to make strategic decisions? Many companies treat market research like an event. We perform a survey/focus group/

ride-along and use whatever we hear as the market truth du jour. Another "event" shows something different, so we go in a different direction. And so on. No one is looking for patterns, trends, and changes over time. No one's got the whole picture. As a result, organizations miss opportunities in normal times and embarrass themselves during crises.

How to Innovate on Purpose will help you remove the roadblocks to innovation and provide you with a step-by-step framework to:

- Align with leadership so your innovation efforts deliver on organization vision and strategy;
- Build team effectiveness and a united sense of purpose by establishing clear roles in the innovation workflow;
- Aggregate existing market knowledge and fill the most urgent knowledge gaps to improve how you serve customers right now;
- Understand the market and respond to it from a whole-experience perspective so you can build value far beyond new product features;
- Look ahead to anticipate market changes and opportunities; and
- React quickly and well in a crisis.

I'll show you how to do this with whatever amount of time, budget, influence, and leadership buy-in you've got. I've implemented all or parts of this process in many of my prior roles and studied organizations that have done something similar—as well as those that haven't. I've also made plenty of mistakes of my own, and I hope I've learned from those too.

Without a workflow to focus efforts and proactively seek out market-driven opportunities for innovation, organizations get stuck in a quagmire of internal enthusiasms and a long list of user-requested product tweaks. Even though well-meaning, this is the worst way to attempt innovation.

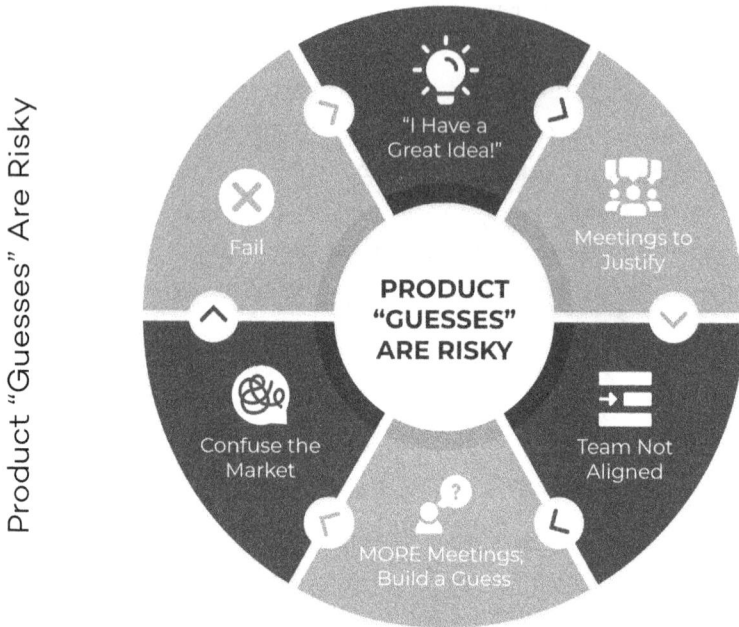

This book won't eliminate surprises, or risks, or wrong turns. What it **should** do is help you better understand your customers, now and as they evolve over time, so you can build products intentionally designed to solve their problems and achieve your own strategic goals.

In the coming chapters, I'll lay out a clear, easy-to-implement five-step workflow you can follow without a lot of extra time or budget to improve your market position with existing products in established products or find that Next Big Thing. This is the workflow I've used, coupled with what I'd improve and add, based on my personal trial (and error), ongoing experience, and research. I'll give you tools, templates, ideas, and anecdotes for each step. At the end of each chapter there'll be recommended next steps for you to follow even if you only have 30 minutes to do them.

Before we get to the workflow, let's get clear on what "Innovation" is.

ACTION PLAN FOR CHAPTER 1:

- Review the Table of Contents of the book and make a list of "how to innovate" questions you hope to answer by reading it.

Chapter 2
WHAT IS "INNOVATION"?

In my experience, very few organizations have a unified idea of what it means to innovate. What's the process for innovation? What's the outcome? Who's in charge of it? Without a clear understanding of what innovation means for your organization, it's nearly impossible to achieve it. So, we'll start here—what is "innovation"? To understand what innovation *is*, let's begin by detailing what it is ***not***.

WHAT INNOVATION IS NOT:

- ***Innovation is not (always) invention.*** Lots of folks confuse innovation with ***invention***. Invention is generally agreed to be the initial creation or discovery of a product or process. But they're not so much products as they are enabling technologies or ways of doing things. For example, cloud computing and mRNA were certainly invented/discovered and then productized, but they're used to improve or even enable innovation in a huge variety of software products and vaccines, respectively. Of course, many innovative new products result from invention, but you don't *have to* invent to innovate.

- ***Innovation is not (only) product.*** Looking beyond product for opportunities to innovate can be a secret weapon for start-ups and established firms alike. As noted in the Sears example, attributes such as pricing, marketing, purchase, and delivery

9

workflows and myriad other non-product innovations could be enough to drive new market behaviors even if the product that market buys is exactly the same. From Sears to Netflix to Starbucks to Lyft, savvy start-ups have taken advantage of 360-degree innovation strategies to disrupt established players while selling the same product they do.

Sometimes innovation isn't even revenue-generating! A mentor of mine in the law firm marketing space implemented a customer portal that allowed dynamic update of the firms' online presence, such as newly hired lawyers or expanded areas of practice, so the firms could market these advantages right away. It wasn't a product, but it did make the product they were buying—law firm visibility—much more valuable.

- *Innovation is not (always) something new.* Taking into account the first two "what innovation is not" bullets, it should be evident that innovation isn't synonymous with only brand-new products. Businesses can continue to innovate in existing markets, leveraging existing products, and likely should. What innovation means to your organization with regard to scope and scale is something we'll touch on very soon, but the point is that you should think of innovation, and get aligned around how you'll innovate, even for mature products and markets.

- *Innovation is not a mystery.* Many folks believe innovation is an internal, personal "lightbulb moment." A mystery—almost a miracle—that occurs to a genius like a brainwave. How could innovation possibly happen by doing something as boring as responding to known market needs with the simplest acceptable solution? This perspective makes it easy to believe that innovation is out of our grasp—hey, if Steve Jobs worked here, we'd be innovating all day long! But innovating on purpose, methodically, and iteratively is what all those geniuses did too—we just don't hear about the blood, sweat, and tears they put into that miracle product we read about in our college classes and *Wired* magazine.

- ***Innovation is not absolute.*** Every industry is different, and within each industry, every organization has a vision and strategy to stand out from the competitors and stay relevant to the markets they want to serve. One organization wants to be the biggest solution provider; another, the very best local option. One wants to be known as the "out there" company, coming up with the wild new products only a handful of people in the market will dare to try. The other, the forward-thinking but still mainstream provider of great offerings. Clearly, the market opportunities those two are searching for wouldn't be the same.

That's a lot of examples of what Innovation is not; let's move on to what innovation ***is***:

WHAT IS INNOVATION?
- ***Innovation is inspired by a market.*** Innovation opportunities come from others needing help or wanting enjoyment in new or better ways than they have now. It may be as simple as hearing them when they say, "Build me this." More often, you'll need techniques to dig deeper; we'll discuss those later in the book.

- ***Innovation is inspired by your organization vision.*** Some organizations set up a big roadblock to purposeful innovation themselves by not defining the markets, technology, or types of problems they want to solve. In other words, they feel like they should look everywhere, all at once, all the time. My advice is: don't. Put some guardrails in place to focus your energy in areas you intend to be good at. Then, go looking for innovation opportunities at whatever level of disruption you aspire to.

- **Innovation is everywhere.** That said, the *inspiration* for innovation can—and should—be found everywhere. You should focus your innovation on markets and types of activities you're good at, but the inspiration for that innovation could come from changing social attitudes, unrelated new technologies, or marketing methods. It could come from regulations that impact you or your markets or supply chain challenges that create a new need. In other words, most organizations need to focus where you look for opportunities but expand where you look for the tools to capitalize on them.

- **Innovation is up to you.** And this leads us to the best working definition of *innovation* for your organization: *it's up to you.* To innovate on purpose, your leadership must define what innovation means at the intersection of what your market needs and who you want to be. How do you do that? It's time to create a common language around *innovation* in your organization.

INNOVATE ON PURPOSE USING THE INNOVATION SPECTRUM

Is the primary goal of your organization to keep existing customers happy with an existing product? Or are you leading the charge to solve a brand-new challenge for a market that didn't even exist last year? Maybe you're pursuing the latter in one part of your organization and the former in another? And maybe—maybe, you're not sure.

Without this high-level alignment on the scope of innovation expected of you, it's pretty hard to come up with ideas leadership will love (and fund)! And practically, it's impossible for anyone to watch every market, all the time, for any possible opportunity. That's why many companies end up chasing too many opportunities with product "guesses" and hoping for the best. We need a way to focus innovation efforts without stifling them.

To innovate on purpose at any organization requires a common understanding of what phase of innovation you aspire to. To help solidify this understanding, I've created **the Innovation Spectrum**. It provides four phases of innovation—Reactive, Responsive, Inventive, and Disruptive—to help teams define how innovative they're expected to be. The intent is to create focus and manage expectations for your innovation efforts, from what market research you do to managing user requests.

The Innovation Spectrum defines four phases of innovation an organization can aspire to. It's used to solidify alignment, create focus, and manage expectations for your innovation efforts throughout the Innovate on Purpose Workflow.

Each phase of the Spectrum illustrates an increasingly more radical and forward-looking type of innovation, which is increasing risky, increasingly rewarding, and increasingly broad in the scope of market research and experimentation you'll need to do to find it. Your goal is to find out what level of innovation is expected from you and your products.

REACTIVE INNOVATION

The goal of **Reactive Innovation** is to continuously fulfill current product users' requests to improve existing solutions. Of the four phases, this one could be arguably said to *not* be innovation, because it's characterized by fulfilling requests instead of doing proactive opportunity identification. In this phase, users tell you what they want—maybe literally draw a picture and say, "Build me this." Organizations serving very mature markets, where there are no new customers to get, or where renewals are driving the lion's share of profits, need to focus at least some effort here. But even when renewals are paramount, some innovation beyond order-taking is a good idea to mitigate the risk of (eventual) obsolescence or disruption. So let's move on to the next step on the Innovation Spectrum: Responsive Innovation.

Innovation Spectrum

INNOVATION SPECTRUM

	REACTIVE 	RESPONSIVE 	INVENTIVE 	DISRUPTIVE 
DEFINITION	Reactive Response	Additive Upgrade	Portfolio Expansion	Opportunistic Diversion
EXAMPLES	Improved Interface Bug Fixes	New Reports New Integrations	New Ecosystem New Pricing Model	New Solutions New Problems Solved
PRINCIPAL DRIVERS	Customer Directive	Market Expectations	Organization Strategy Market Anticipation	Trends Non-Linear Synthesis
BENEFITS	Customer Intimacy	Share of Wallet	Market Expansion	Industry Leadership Singularity
CHALLENGES	Prioritization Disruption	Proactive Outreach Adoption	Focus Awareness	Scale Funding
RISK/REWARD	High/Low	Low/Medium	Medium/High	High/Very High

RESPONSIVE INNOVATION

Responsive Innovation allows you to find current-market opportunities to expand functionality of existing products that can enable you to increase user *and* buyer satisfaction (and, ideally, renewals), steal market share from your competitor, or even bring in new customers from the markets you currently reach. It helps you know when it's time to modernize your product to upgrade security, interoperability, and scalability. Responsive Innovation goes beyond what users ask for to proactively innovating to satisfy buyer requirements, upcoming changes in regulatory requirements, or to stay ahead of competitors.

INVENTIVE INNOVATION

Inventive Innovation is about building better solutions for known market problems or selling existing solutions in new ways, to expand into new markets or customer workflows. Inventive Innovation may be non-product oriented, in the form of a new pricing model, such as offering monthly subscriptions on formerly contract-only products; or new delivery model (grocery delivery, anyone?). Expanding your offering from a product to a workflow or persona ecosystem is another facet to Inventive Innovation. These opportunities are harder to identify in the market, because they require investigation of an overall industry, workflow, and the changes that create such opportunities. It also requires listening for non-product frustrations, such as dissatisfaction with the current distribution, pricing, or selling systems. That said, you can often repurpose all or pieces of mature products in new ways to pursue these opportunities. Inventive Innovation is riskier than the prior two types, but the rewards are significantly higher.

DISRUPTIVE INNOVATION

Disruptive Innovation is what most of us think of when we hear "The Next Big Thing," miracle products that come seemingly out of nowhere to solve problems we didn't even know we had. That's not how it happens, but the goal of Disruptive Innovation is indeed to articulate and solve market problems that were previously unarticulated, unsolved, or nonexistent. These opportunities can emerge as a result of invention or improvement of enabling technology, a cultural re-prioritization of needs, a shift in values, or the emergence of a new market entirely.

Let's look at an example of each phase. Let's say your organization offers time-and-billing software to help law firms track the amount of time each lawyer spends on a certain client's case so they can charge those clients appropriately. What type of innovation opportunity does leadership want you looking for?

- Improve your existing software to populate the billing code wherever needed = **Reactive Innovation.**
- Add the ability to report hours worked by practice area because accounting regulations will require this in two years' time = **Responsive Innovation.**
- Offer your software on a month-to-month subscription with a freemium version for solo law practitioners = **Inventive Innovation.**
- Leverage the data flowing through the software to model legal billing trends; create an app for corporate *clients* of legal services, to help them forecast and negotiate their legal fees in a way not available today = **Disruptive Innovation.**

Talking through such order-of-magnitude examples can guide you to a common understanding of the type of innovation opportunity you should look for. Budgets, resources, timelines, and risk all escalate the farther down the spectrum you go, as does the opportunity for reward, competitive dominance, and market disruption. Not everyone can or should be looking for Disruptive Innovation, but if you are, everyone must be clear on what that entails: broad, detailed market research, and a continuous, rapid cycle of experiments to validate learnings.

With scarce resources, fixed budgets, and never enough time to do everything, getting straight on what leadership wants the business to accomplish is the first step toward successful innovation.

OTHER CONSIDERATIONS TO DEFINE INNOVATION

Businesses exist for the purpose of serving markets, employees, and owners (I'll leave it to others to debate the order those should appear in); which means there are guidelines any business, no matter how driven by humanitarian- or mission-related principles instead of profit or revenue, must keep in mind when considering innovation opportunities:

- Innovation is the solving of existing problems in better ways, the solving of existing problems that were previously unsolvable, or the identification and solving of previously undefined problems. *To be considered innovation, the effort must solve a market problem someone is willing and able to pay to solve.*

- Another market-driven requirement of innovation is that it be usable. *To be considered innovation, the effort must solve a market problem in a way the market is capable of and willing to adopt.*

- The reason organizations innovate is to serve markets they're qualified to serve. For a business, innovation is focused on what your organization aspires to do and who you aspire to do it for. *To be considered innovation, the effort must contribute to driving your own organizational financial goals, vision, and strategy.*

- All of the above must be understood by those in your organization who are tasked with identifying innovation opportunities. *To drive innovation, your organization must have a clear definition of what that means in terms of target markets, strategic alignment, and appetite for scope, scale, budgets, and timeline—as well as who owns the innovation workflow.*

WHO'S RESPONSIBLE FOR INNOVATION?

Most organizations are spending millions on market research yet still find it nearly impossible to deliver innovative new solutions their markets want to buy. In others, product and development teams conscientiously fulfill what users ask for in release after release—only to be thrown over by those same users for some radical start-up offering. Why?

These challenges are due to a critical ownership gap. In most organizations, there is no one in charge of deciding what questions need answering. No one to say what personas and markets to focus on and how radically unique the innovation you go looking for should be. No one leading brainstorming on innovation options such as pricing, messaging, sales channels, or interoperability in coordination with product and commercialization enhancements.

There's no Market Strategist.

WHAT IS A "MARKET STRATEGIST?"

The **Market Strategist** is responsible for driving the workflow that enables purposeful innovation. This person establishes the scope and priorities for market research to deliver efficient but broad insight to inspire innovation. They analyze aggregated insights to identify and prioritize market-driven innovation opportunities aligned with their organization's strategic direction. They communicate the priorities and the context that drove them to enable teams to design, price, sell, and support solutions that delight markets, disrupt competition, and solve new or previously unsolvable problems. They're responsible for these activities whether the opportunity to innovate comes in the form of a better product, improved delivery, powerful marketing, or smart pricing. They're responsible for connecting the dots between buyers and users, the market and corporate goals, and emerging technologies and tools to recommend the Right Ideas for innovation.

Market Strategist: the individual (or group) responsible for crafting a market research and opportunity identification strategy that aligns with organizational goals and delivers market-driven innovation opportunities. Ideally this is an acknowledged role, but it can be combined with the responsibilities of the product manager or product marketing manager.

I recommend hiring one or several individuals to perform this function, ideally elevated from your current product management or commercialization team. This person would report to the CEO and manage the product management and product marketing teams; they could also manage the UX and data science team depending on the type of products you build and how tightly these teams would work together. Each Market Strategist could be responsible for a market (e.g., "law firms in the US"), a function (e.g., "marketing solutions for US law firms"), or a segment that makes sense in your world.

Organizations get the most value from the Innovate on Purpose Workflow if you have someone to lead it, but that said, it's much more important to ensure that someone—whatever their title—is responsible for *executing* the activities associated with being a Market Strategist. So even if you can't make the budget and organization decisions to create this role, I'll show you how to implement the pieces under your purview.

WHAT DOES THE MARKET STRATEGIST DO?

The Market Strategist finds innovation opportunities in target markets, tempered with and focused by the stated strategy of their own organization and the opportunities to serve both. They aggregate existing market information, fill the gaps they discover, and empower their teams with the market knowledge to design, build, price, and sell products those markets will love.

KEY RESPONSIBILITIES OF THE MARKET STRATEGIST

The role of the Market Strategist is focused on the following:

- *Understand what innovation means to their leadership.* "Innovation" doesn't always mean radical, disruptive new offerings. For some organizations, it means serving existing customers in new ways or offering existing products in new markets. The Market Strategist gets aligned with leadership on the scope and degree of innovation desired to focus team efforts at the right level.

- *Establish a knowledge-gathering workflow that will result in market-driven innovation.* This includes establishing a baseline of current market knowledge, prioritizing filling the gaps in that current knowledge, and establishing ongoing workflows designed to ensure the team can act quickly if something changes. The Strategist listens beyond buyers and users to find, for example, new ways to sell products, changing industry regulations, and emerging enabling technology. They also listen to their colleagues, who often are the direct line of communication and expertise, and include their insights in the overall market view.

- *Create a language of understanding.* Most organizations don't have a unified vocabulary for insight confidence—an answer to the question "How sure are we that this is true?" The Market Strategist creates confidence thresholds and a vocabulary, ensuring their colleagues understand the risk in the knowledge they're using to make decisions.

- *Focus and align teams.* The Market Strategist focuses and aligns teams in two key ways. First, by focusing everyone to listen and observe with strategic goals in mind. The Market Strategist is in no way the only one listening to the market! Ideas will come from everywhere, and it's the Market Strategist's role to make sure the team brings insights and ideas that are aligned with strategy and the level of innovation the organization aspires to. On the back end, Market Strategists find effective ways to disseminate market and corporate knowledge to internal teams so they can do their jobs better. The Strategist will certainly participate in ideation and prioritization sessions (perhaps even lead them) but will find other channels to layer the knowledge into the organization so everyone from the product manager to the contracts lawyer can do their jobs well.

- *Set an example for innovating on purpose.* The Market Strategist is in a unique position to curtail well-meant but ineffective "innovation by guesswork" activities that teams often do in the absence of market knowledge. Wherever possible, the Strategist should bring market and strategy knowledge to these discussions and challenge teams to experiment and validate prior to building full-blown offerings for the market.

Listening to the market is something all organizations spend money and effort to do. Unfortunately, they seldom have a system in place to focus the efforts, to distribute the learning, or to act on what they gather. The result is a lot of market research that isn't particularly valuable and a small amount of research that could be valuable but is unknown or ignored. To be wise and effective when listening to the market, someone must be in charge. Adding—or formalizing—the role of Market Strategist is the first step in achieving market-driven innovation.

WHAT IF WE DON'T HAVE A MARKET STRATEGIST?

Most organizations don't have a Market Strategist today. If you're in a position to formalize that role (or add several, if you're a large company with multiple products and markets), terrific. You can often find good candidates in your product management, product marketing, or even UX teams because they're already doing part of the job being the experts on users and buyers. If not? I recommend using the workflow and way of thinking I'll lay out in this book to expand your effectiveness in the role you currently have. I believe this book makes a good case for this role, but it's more important to get the work done than to worry about getting an added headcount or title approved.

Without a workflow to deliver market-driven innovation, an organization is guessing—a risky, wasteful, and frustrating alternative. There's a better way, and we'll talk about that now.

ACTION PLAN FOR CHAPTER 2:

- Practice using the Innovation Spectrum.
    - Print the Innovation Spectrum and review the descriptions of each phase.
    - List the products you're responsible for under the innovation phase you're using to guide your work today. Include services, hardware, software, or internal products if that's your role.
    - Discuss and edit with your leader.

- Determine what Market Strategist responsibilities you're responsible for.
    - Using the Market Strategist role description in this chapter, list which of the activities you believe you're responsible for and which you're not. Validate with your leader.
    - Or, if it's within your purview, create a Market Strategist job description from the information in this chapter and post it for hire.

Chapter 3
CREATING A WORKFLOW TO INNOVATE—OVERVIEW

Most organizations are, in fact, listening to their markets. Some are doing it quite well, and quite thoroughly—but nearly all perform market research by rote, without any conscious alignment with strategy or intentional searching out of opportunities to innovate. Further, they often fail to communicate, heed, or leverage what they learn. In pursuit of innovation, teams often throw a lot of solid market knowledge out the window and rely on the ideas they come up on the fly.

To deliver opportunities for innovation that both drive success in the market and deliver on your organization's strategic goals, we need an intentional workflow around opportunity discovery: a workflow to innovate on purpose.

The goal of this workflow is to transform your current market research and project prioritization from haphazard guesswork to a continuous, strategic engagement with your markets and effective, purposeful innovation. It will enable you to:

- Respond to your organization's vision and strategy when prioritizing where you go looking for innovation opportunities;
- Enable a complete view of the market and opportunities through aggregation and synthesis;
- Focus and align teams by creating a language of understanding;
- Use what you already know to win in the market; and
- Be ready to act when the market changes.

For ease of learning, the workflow is laid out in linear steps with discrete activities, tools, and deliverables tied to each. In the wild, your own workflow will more likely resemble an ongoing cycle to continuously hone and improve each step. You may need to make a decision now for Step 1 that would benefit from the learnings of Step 3. Over time, you'll gather those learnings and revise plans and assumptions made earlier. As a matter of fact, we'll be creating a Confidence Language along the way to help everyone understand how much actual insight any given decision is based on. So learn the concepts and leverage the Action Plan at the end of each chapter; then go back and make each step better over time.

FIVE STEPS TO INNOVATE ON PURPOSE

The path to innovation isn't guessing and hoping. To deliver market-driven innovation that also drives your organization's success, you need to be methodical. At a high level, there are five steps to creating a plan for market-driven innovation: Strategic Planning, Organizing the Work, Finding Opportunities, Empowering Teams, and Responding to Change. Under many of these steps are specific activities that warrant discussion as well.

Here in Chapter 3, I'll give you a high-level summary of each step and, where applicable, an activity. The following chapters will give you the details and tools to complete each one.

FIVE STEPS TO INNOVATE ON PURPOSE

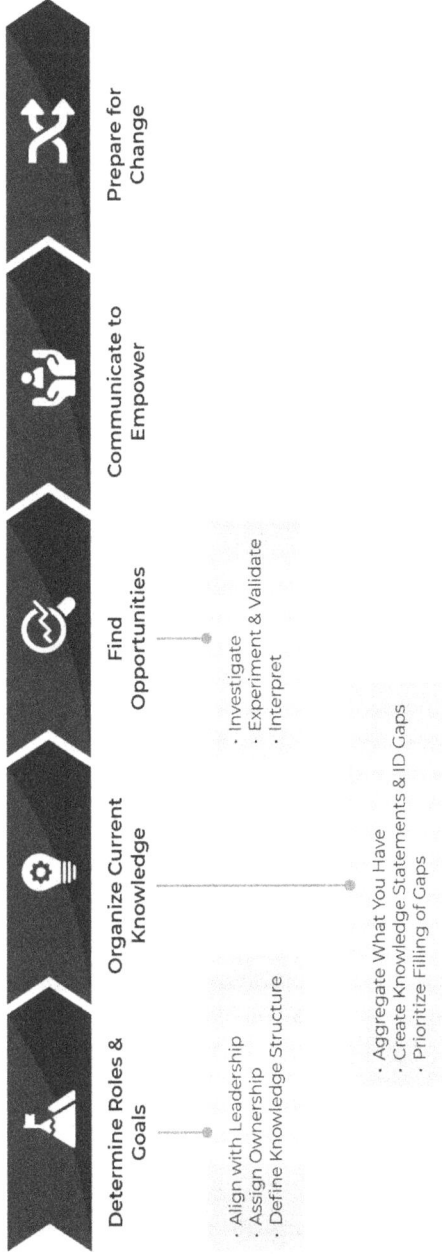

Determine Roles & Goals

- Align with Leadership
- Assign Ownership
- Define Knowledge Structure

Organize Current Knowledge

- Aggregate What You Have
- Create Knowledge Statements & ID Gaps
- Prioritize Filling of Gaps

Find Opportunities

- Investigate
- Experiment & Validate
- Interpret

Communicate to Empower

Prepare for Change

Five Steps to Innovate on Purpose

STEP #1: DETERMINE ROLES AND GOALS

This is a level-setting step to kick off your workflow, even if you don't have all the knowledge you'd like to have to deliver excellent results. The purpose is to point you in the right direction and clarify with leadership who's on the team and what they're responsible for. This step will be broken down into three chapters, as follows:

- *Align with leadership.* The Market Strategist is a strategic partner to organizational leadership, and the first step to innovating on purpose is to ensure that the work done will drive leadership's goals for the organization. With scarce resources, fixed budgets, and never enough time to do everything, getting straight on what leadership wants the business to accomplish is the first step toward prioritizing your work.

- *Assign ownership.* Determine who **owns** the process and who **feeds** the process. In this section I'll offer specific tools to clarify roles, enable team members to perform them, and get leadership on board with the contributions needed from each team to succeed.

- *Define a knowledge structure.* A common language will be necessary to align and empower your organization to innovate. In this step I'll offer a perspective on taxonomies or data structures to structure your information. I'll detail the information that's needed to enable innovation, whether you use a sophisticated app or a basic spreadsheet. We'll also agree on a set of thresholds that will communicate to the organization how "true" the information they're using to innovate actually is.

STEP #2: ORGANIZE CURRENT KNOWLEDGE

In this step, you'll gather up the market knowledge that already exists, compare it to strategic goals, and identify gaps to fill while you continue to innovate based on the current knowledge. This step will also be broken into three chapters, detailing the following activities:

- *Aggregate what you have.* Collect all the market research and accepted insights your company has gathered in the past. Take advantage of the work you've already done to drive your immediate next steps. You already know more than you think you do!

- *Create knowledge statements and identify gaps.* We'll walk through how to perform a systematic review of the data you already have in-house and crystallize it into actionable statements. This part of the workflow will make it easier to determine where you need to direct your ongoing research and make every market research dollar count.

- *Prioritize filling knowledge gaps.* You'll never have the time, budget, and resources to keep current on all possible opportunities for innovation–much less execute on all of them. This section will guide you to make the right decisions for your company and align your team around them. It will also offer ideas on creating a market research backlog.

You'll have lots of great opportunities after this work is done, but you'll need to look to the future as well. The next step begins that work.

STEP #3: FIND OPPORTUNITIES

Now it's time to fill the critical remaining knowledge gaps and find new opportunities for innovation. Too many organizations think this phase is full of magic and miracles–something that can only happen if you have a resident genius and crazy-good luck. Not so. Activities in this step include:

- *Investigate.* Often teams believe something to be true and stop there. When creating an Innovation Map, you will identify the knowledge gaps you have and hypothesize what you believe to be true. This chapter will give you some techniques to do that.

- *Hypothesize and validate.* Quantify and verify that you know what you think you know. How many? How much? How badly? How often? These are the elements you'll validate, and in this chapter we'll talk about some best practices to do so.

- *Interpret.* This chapter is where you bring your big brain to the table. We'll discuss methods to bring your validated market insights together to identify potential innovation opportunities. It's not about miracles or magic—we're going to work the data.

But you're not the only one whose big brain will be needed to succeed. To innovate on purpose means getting the whole team engaged. For them to shine at maximum brilliance, they have to have the same insights you do.

STEP #4: COMMUNICATE TO EMPOWER

Market knowledge is only as valuable as your response to the opportunity you find. To maximize your results, everyone in the organization must have a thorough and current understanding of your market and strategic goals. But how? In this chapter, I'll provide some best practices and templates to get all this great information to your cross-functional team, while minimizing the time and resources required to do so.

You're aligned with leadership. You're using what you already know to innovate on purpose. You're filling the knowledge gaps, expanding your insights, and brainstorming 360-degree innovation options with the team. So we're done! Right?

STEP #5: PREPARE FOR CHANGE

Wrong. Most organizations perform market research as isolated tactics that produce a single result. Do a focus group, report the results; create a persona, revise it in two years. But markets change, regulations change, the people doing the jobs change—sometimes slowly over time but occasionally quickly, radically, and significantly (COVID-19, anyone?). This step extends the best practices you've learned to discover innovation opportunities into a workflow to listen for and manage change. This workflow will help you handle surprises big and small so you can respond quickly and effectively, even in the most unexpected situations.

It's time to dive into each step of the Market Strategist's workflow to innovate on purpose.

ACTION PLAN FOR CHAPTER 3

- Print the *5 Steps to Innovate on Purpose* graphic to use as a reference while you read.
- Quickly scan each chapter of the book and highlight the ones you're most in need of on your printed *5 Steps*. It's good to refer back to the questions you wrote down after reading Chapter 1.
- Skim the chapters you already have a process for, or aren't in charge of, in order to spend time on the chapters and steps that will help you most.

Chapter 4
ALIGN WITH LEADERSHIP

Determine Roles & Goals

- Align with Leadership
- Assign Ownership
- Define Knowledge Structure

Have you ever worked for a CEO who said they wanted to see every idea but rejected 99% of them? Or reported to someone who hired you to lead innovation but was guided by the opinions of long-term employees? Perhaps you've submitted business plans full of market insights, only to see leadership's gut instinct get the focus and the funding. These issues are real and will never be completely eliminated, but you can mitigate them by creating alignment with your leadership on the form and direction market-driven innovation should take.

Some of you may already have a clear idea how you fit in to, and contribute to achieving, your organization's vision and strategy. But too often, Market Strategists skip the critical step of aligning with leadership on the organization's market and product focus, appetite for innovation and timeline, and scale of expected results. Without this alignment, you risk spending a lot of time and effort writing proposals and doing market research your organization will never use, frustrating all concerned. The good news is that a discussion leadership, structured around a few simple questions, should provide what you need to deliver market-driven innovation that *also* drives organizational strategy.

In this chapter, I'll cover several areas to get aligned with leadership on, and the questions to help you have the conversations. I'll also provide some ways to visualize the insights you gather and a workflow to create your own contributing strategy as a response to the corporate strategy.

Let's start with the topics on which to get aligned with leadership.

ALIGN ON QUICK WINS AND TOP-OF-MIND CHALLENGES

When you start the conversation about aligning your work with corporate strategy, you may hear some pent-up desire from leadership to pursue a few specific opportunities. Maybe your R&D team has come up with something your execs would like to commercialize quickly. Perhaps regulatory changes have been announced but not reviewed to determine whether product changes should be made. It may be a pet peeve your leaders realize know isn't the top concern in the market but still bugs them enough that they won't stop thinking about it until it's resolved.

These top-of-mind challenges may not be foundational to your organization's vision and strategy. They may not be the opportunities you expected to address. But listen for them, as they may be quick wins to build trust and gain the engagement of your leaders.

You may also have a few quick wins you can *suggest* to leadership. Several years ago, I came in to lead a small division of a larger company and discovered no one was following up on free product trials. Some of the "trials" had been going on for 18 months! It was easy to shift the sales team's focus from trials to closing deals and brought in revenue the division needed to gain credibility with leadership.

These possible quick wins may be small impact, or based more on opinions and hopes than market knowledge, but resolving them brings the team together and provides a short-term demonstration of value while you investigate strategic, market-driven, or longer-term innovation opportunities. So have a quick brainstorming session to identify these projects, if they exist. Some questions you could ask to get there include:

- Is there a specific product, market, persona, or competitor you are actively concerned about right now? What's your main concern?
- Is there a stalled opportunity that you'd like tackled right away?
- For these concerns and opportunities, what is the most important information we *should* have on this [market/persona/competitor] that we *don't* have?
- What have been the consequences of *not* acting on these items?
- What are the anticipated benefits of delivering on these efforts?
- What does "resolved" look like to you?
- How much budget and time do you want the team to spend on resolving these items, relative to diving into creating a long-term innovation workflow?

Have this conversation first. Quick wins are what you do while you lay the foundation for the harder work—ongoing improvement to your innovation process and results.

What if the project isn't a quick win? What if it would take significant time and resources? Include it in your backlog—something I'll talk about later—and we'll prioritize it with ongoing projects.

Let's move on to the longer-term alignment with leadership, starting with understanding what drives your corporate vision and strategy.

ALIGN ON THE CORPORATE DRIVERS OF INNOVATION

As the Market Strategist, where in the market should you be looking for opportunities for market-driven innovation?

It's a big world out there, and it's impossible to search for innovation opportunities in every market, for any type of solution, everywhere, all the time. The good news is, leadership doesn't want or expect this, no matter how broad the rhetoric is around finding that Next Big Thing. Having a working knowledge of organizational vision and strategy is the first step to align with leadership and focus your efforts. If you already have that, great. If not, read on:

UNDERSTANDING THE CORPORATE VISION AND STRATEGY

Understanding the overarching direction of your organization will also help you be seen as a strategic asset, even if your personal focus is on achieving a small portion of the corporate goals. Getting to alignment and understanding of the organization-wide or **corporate vision and strategy** is a matter of a few steps, starting with understanding what corporate vision and strategy *are*.

> **Corporate Vision & Strategy: Statements to express what your company aspires to be in the markets they serve and the very high-level actions they'll take to get there.**

A statement of your corporate *vision* is an aspirational description of what your organization wants to ***do for*** and ***mean to*** the markets they'll serve in the future. It's like today's casual runner saying they want to compete in marathons in the future.

Strategy lays out, at a very high level, the actions you'll take to achieve your vision. For the runner, it might be adding strength training and increasing the distance of their training runs.

Here's an example of a company vision and strategy statement for an imaginary company we'll call LegalcoEG, a company that builds workflow-enablement software for law firms:

> *As the premier provider of time-and-billing solutions for South American law firms, our **vision** is to help law firms everywhere increase their effectiveness, efficiency, and profitability through better workflow enablement tools as the go-to company for the most sophisticated solutions for running a law firm.*
>
> *Our **strategy** to achieve this includes expanding our product line to offer end-to-end workflow solutions, first for law firm finance departments and then for all aspects of running a law firm. We'll enter the Canadian and US markets while assessing next steps for further globalization. We'll expand our solutions beyond software to*

*integration, implementation, best-practices consulting, and other offerings
to become the trusted back-office partner for global law firms.*

Your corporate vision and strategy may give you all the
understanding you need of how and where your company wants to
innovate in the future. But what if it doesn't? Following are questions
you can ask leadership to align on where the company is trying to go
over time.

WHO ARE WE TODAY?

First, a question that may seem obvious: Who are you today? The
vision statement shows you who you **want to** be, but you'd be
surprised how few organizations formally articulate who they **are
today** in a way that creates an actionable baseline for teams to work
from—a big gap if you want to move forward from where you are!

Even if everyone is aligned on what you do in general, you as the
Market Strategist need a specific and straightforward definition to
establish a baseline for innovation. A good way to begin is to articulate
these four aspects of your organization:

- *Who do we serve?* These are the people who buy and use your
  products and could be expressed in terms of demographics
  (men between the ages of 45 and 65 with no college degree),
  industry (window manufacturers), function (accountants),
  geography (central Texas) or whatever works to focus your
  efforts on one or a few groups.

- *What problems do we solve for the groups we serve?* Most
  companies don't solve every problem for a market, even if they
  solve many of them. Maybe you offer continuing education to
  accountants so they can keep current with the latest regulations;
  maybe you build software that allows them to score outstanding
  invoices by risk of non-payment.

- *What kinds of products do we create to solve those problems?* What do you build to solve these problems? At the highest level, it could be software, data, scoring models, coursework, services, or hard goods.

- *What is our unique market position?* Said another way, where do we fit in with the alternative ways our market could solve this type of problem? Are you the most trusted provider? The budget alternative?

To help this conversation along, you can create a visual to represent what you *think* is right and verify with leadership. Here's an example for LegalcoEG:

DEFINING YOUR ORGANIZATION: WHO IS LegalcoEG?
Establishing a baseline of who you are, by answering four key questions

PROBLEMS WE SOLVE
Improve law firm cash flow by ensuring clients are billed in a timely manner.

PRODUCTS WE BUILD
Billing software

MARKETS WE SERVE
Large law firms in South America

WHERE WE FIT
Trusted provider of a sophisticated, but specific, workflow enablement solution

LegalcoEG is currently positioned as the trusted provider of time-and-billing software for large law firms in South America. This is who LegalcoEG is today. The next step is understanding who you're trying to become.

WHO DO YOU ASPIRE TO BE? CORPORATE VISION STATEMENT

Hopefully, your organization already has a vision statement. If not, craft one in the same way you did the present-state positioning, changing the "What We Do" to "What We WILL Do" in each category, as shown in the example below:

WHO YOU ASPIRE TO BE: DEFINING LegalcoEG'S CORPORATE VISION
Establishing a baseline of who you are, by answering four key questions

Who You Aspire to Be: Defining LegalcoEG's Corporate Vision

If you compare LegalcoEG's "Product We Build" to "Products We WILL Build," you can see that they want to expand their product line beyond software to include the data, analytics, implementation, and consulting needed for a full-service solution. They have other big plans too: it looks like they're going to expand into enabling other back-office workflows for law firms, and expand that law-firm market focus from South America to global reach.

HOW DO WE PLAN TO GET THERE? CORPORATE STRATEGY

Contrasting who you are today with who you want to become raises the next question: How does leadership envision getting from who we are today to who we want to be? In other words, corporate strategy.

HOW YOU'LL GET THERE: DEFINING LegalcoEG'S STRATEGY
Strategy is how you'll get from the baseline to the vision

How You'll Get There: Defining LegalcoEG's Strategy

The answer should give you a sense of what the organization is willing to change, update, and expand in order to achieve the vision. In our example above, LegalcoEG wants to become a global company but has made it clear in the strategy that the very next steps beyond South America will be to Canada and then the US.

Even without a written corporate strategy, you can still reach alignment with leadership by asking these questions.

Is anyone thinking, "That's great, but we can't just keep piling on more work without more resources." I agree! And during the alignment step we're trying to get focused and spend resources wisely, so a discussion of balance is next on the list.

WHAT ARE THE TRADE-OFFS TO GET THERE?

Another point of clarity that should emerge from this discussion are necessary shifts in day-to-day priorities and focus to work the corporate strategy and achieve the corporate vision. Perhaps your current product pipeline and market research are driving an old strategy—or weren't ever driving the strategy at all.

But as you look at the changes shown in the Today/Tomorrow visual below, it's good to follow on by discussing at a high level where more—and less—tactical focus should be applied as you work toward achieving the corporate vision.

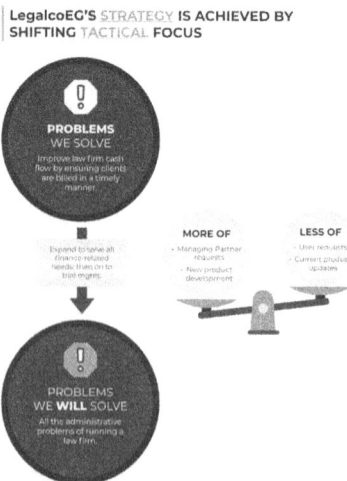

LegalcoEG's Strategy is Achieved by Shifting Tactical Focus

LegalcoEG'S STRATEGY IS ACHIEVED BY SHIFTING TACTICAL FOCUS

So for each aspect of the vision, I recommend asking:

- In order to move from who we are today to the vision of who we will be, what would you expect us to spend **more** time and resources on?
- What would you expect us to spend **less** on?

In our LegalcoEG example, leadership expects the product teams to build new products to expand farther into their customers' finance departments. Then, on to offerings used directly by the lawyers in case and trial management. This means finding out what the most pressing emerging needs of the law firms are and building solutions.

The South American users of the current product aren't the right people to ask about what their leadership needs in other parts of the law firm, nor are they the ones who can tell us what their peers in Ontario or Oslo need.

But wait—there are a lot of current-user requests in the pipeline for the existing time-and-billing software tool! And there are quite a few enhancements you've been itching to include based on a recent focus group. You should do those too, right? The short answer is, maybe not.

To do more in some areas, you'll need to do less in others. Based on the corporate strategy, you may need to focus less on those tactics and more on interviewing the managing partners of your law firm customers about the emerging or unmet needs that they have and start talking to potential product users in Canada.

Strategic goals are achieved in large part through tactical execution, and the evolution of a company from one position in the market to another requires trade-offs. This includes doing more of some tactics and less of others. Leadership should be able to answer this question at an organizational level.

CORPORATE DRIVERS PROVIDE THE CONTEXT TO INNOVATE ON PURPOSE

Corporate drivers are the broad context you need to innovate on purpose. But what if you're the Market Strategist for a single product in a broad portfolio, instead of for the entire company?

Sometimes, a Market Strategist is one of a team of strategists, each responsible for a portfolio of products, a particular market or some other subset of the overall organization. If that's the situation you're in, your job is to deliver on the promise of the corporate vision and strategy through market research and innovation decisions within your scope of responsibility.

It's time to turn our attention to your role in achieving the corporate vision and strategy: your contributing strategy.

FROM CORPORATE TO CONTRIBUTING STRATEGY

Most organizations spend a lot of time and money honing their corporate vision and strategy statements, the direction that guides the entire company. Unfortunately, larger companies often don't validate with product teams exactly how each one plans to contribute to achieving the vision. Without this explicit tie, it's really difficult for individual contributors on any team to self-direct their priorities or focus their efforts. In other words, in a company with multiple Market Strategists, you can't achieve the corporate vision without crafting one or many **contributing strategies.**

A contributing strategy fits underneath the corporate strategy and lays out how one product group contributes (again, at a very high level) to the overall direction the company's headed. In very large organizations, you may have dozens of contributing strategies; in smaller concerns, maybe just a few.

> **Contributing Strategy: Created by the Market Strategist, this statement provides high-level insight into how their specific product responsibilities deliver on the corporate strategy, to help achieve the corporate vision. A contributing strategy**

aligns the day-to-day work of a cross-functional product team. Generally, you'll need one contributing strategy for every Market Strategist.

Why do you need contributing strategies? Because different parts of the organization will help achieve the corporate vision in different ways. Some may be responsible for keeping customers happy with legacy products while others build new products and expand into new markets. Some may contribute to achieving market dominance while others improve product quality. The contributing strategy also helps build a bridge in larger companies from the corporate vision to the work, priorities, and focus of individual team members in their day-to-day workflow. Without the next level of understanding provided by the contributing strategy, it might be hard for individuals in large companies to figure out what to do—and therefore they may try to do everything or focus on a direction counter to the intentions of leadership.

How do you know how many contributing strategies are the right number? A good way to think about it is the way the teams are organized and/or incentivized. For example, let's say your organization builds workflow-enablement software for law firms. It's a big company, and you are the Market Strategist specifically responsible for the software lawyers use to organize their exhibits for trial. Another team, with another Strategist, is responsible for scheduling software solutions and a third for billing tools. There could be many products under each of these teams, or just one. In this example, you should have a corporate vision and strategy, and three contributing strategies, one for each product group.

How many contributing strategies and which products or markets sit within them will be a decision for your leadership, so you need to get aligned on where *you* fit and what *you're* responsible for before building one out. We'll turn to that now: How do you move from understanding corporate vision and strategy to creating your contributing strategy to achieve it? Start with aligning on your scope of responsibility.

ALIGN ON YOUR SCOPE OF RESPONSIBILITY

If you're the Chief Market Strategist, you help craft the overall corporate vision and strategy. If you're a subordinate Market Strategist, you're responsible for delivering on some portion of that vision, as directed by the strategy, and that's where we'll go next. Now that you have some clarity on vision and strategy at the *organizational* level, it's time to think about how *your work* contributes to achieving them.

To do this, discuss the four organizational aspects we used earlier to clarify what your organization does with your leader, only this time talk about what *your* subset of the overall pie is, as shown in the graphic below:

Defining Your Role in Achieving Company Goals

DEFINING YOUR ROLE IN ACHIEVING COMPANY GOALS
Understanding your strategic responsibilities is the first step toward tactical effectiveness.

Do YOU focus on software, services, or solutions?

How do YOU reinforce your market position?

PROBLEMS WE SOLVE
How do we make the lives of our customers better?

PRODUCTS WE BUILD
What combination of hard goods, services, consumables, or software do we build?

MARKETS WE SERVE
Industries, roles, demographics, or other way to establish who we exchange value with?

WHERE WE FIT
What's our market position - budget provider, custom solution, or something else?

What market problems are YOU responsible for solving?

Maybe YOU focus on law firms while others help universities?

Now that you understand what you're responsible for, you can start building your contributing strategy that answers the same questions as the corporate strategy. You'll address:

- What parts of who we are as a company are you responsible for?
- What parts of the corporate vision are you expected to deliver? What is the corporate strategy for each of those?

- What is your contributing strategy? Using the corporate strategy as a guide, create a high-level set of actions that take your areas of responsibility from who we are today to who we aspire to become.
- Identify key actions and areas of focus where you'll do more, and where you'll do less, in order to ensure that scarce resources are focused on achieving the corporate vision.

Using the framework of LegalcoEG's corporate vision and strategy from earlier, the figure below illustrates specifically what you're responsible for.

Currently LegalcoEG has only one product: time-and-billing software for South American law firms. You are the Market Strategist in charge of the current product. However, now that the company aspires to grow not only into new markets but to build new products, you won't work on every product the company builds.

That said, you're responsible for enabling the current product to meet the needs of customers outside South America, as well as ensuring that the software interoperates with the other offerings LegalcoEG will build in the future.

FROM CORPORATE STRATEGY TO CONTRIBUTING STRATEGY
Define how your responsibilities should contribute to achieving corporate goals

Understanding the *scope* of your responsibilities is the first step toward creating a contributing strategy. The next? Aligning with leadership on the *scale*.

ALIGN ON THE *SCALE* OF YOUR RESPONSIBILITY

Now that you understand the scope of your strategic responsibility, it's time to align on the *scale* of that responsibility. This is the difference between what you're **responsible** for and the **span of options** you have to fulfill those responsibilities. In this book, I'll break that scale down into degree of innovation, level of risk, and volume of committed resources.

Degree of innovation is the first element of scale.

SCALE OF INNOVATION: WHAT A "BIG IDEA" LOOKS LIKE

Execs often bring me in to help teams identify the next Big Idea innovation: the cool new thing that will amaze the market and take the company in a bold new direction. What I've learned is, *my* idea of a Big Idea, the *team's* idea of a Big Idea and *the CEO's* idea of a Big Idea may be completely different. In other words, no one knows what "innovation" *means*. **To innovate on purpose, this has to change.**

Many CEOs I've worked with say, "My team comes up with terrific product *enhancements* all the time but can't find the Big Ideas we need to grow—why not?" On the other hand, I've seen product teams bring fresh new idea after idea, only to get shot down each time. Even with a clear vision and strategy, why is it so tough to find the *right* level of innovation?

On its face, this is a reasonable question. Product teams know their markets, have the best handle on shifting needs, and were hired to develop innovative responses to meet those needs. Yet, while their brainstorming sessions yield good incremental projects, they're seldom the game-changing Big Ideas that could significantly expand the scale and market position of the organization. Frequently, the big ideas teams bring to the table are unaligned with corporate strategy: they're *too big*—not too small. These product teams build proposals, business plans, and personal enthusiasm around these ideas, which for legitimate but unknown-to-them reasons won't be seriously considered by leadership. Leaders get frustrated with lack of progress, and the team feels defeated. Why is getting to the Next Big Thing so difficult? Three words: lack of alignment. Specifically, lack of alignment on what we all mean by a "Big Idea."

To create this alignment, I'm going to take you back to the Innovation Spectrum, introduced in Chapter 1. Using this spectrum as a platform for the Big Idea discussion helps you align with leadership around your area of focus in the market.

Start by clarifying what phase on the Innovation Spectrum your leadership expects from you. Here's how that might go at LegalcoEG. We know LegalcoEG currently creates time-and-billing software to help lawyers at law firms track their work time on cases, so their finance department can bill clients appropriately. What does a Big Idea look like to their leadership? If I were the Market Strategist at LegalcoEG, I'd brainstorm some order-of-magnitude examples with my leadership to reach a common understanding.

- Update reports offered in the existing product to cross-populate more fields, as requested by several current users of the LegalcoEG product. **Reactive Innovation.**
- Add the ability to report hours worked by practice area, which is something LegalcoEG has heard managing partners ask for. **Responsive Innovation.**
- Offer predictive forecasting of likely future billings based on prior work patterns, something observation of management workflows has shown law firms trying to do manually. **Inventive Innovation.**
- Create an app that each attorney installs on their wearable, that uses IoT technology to attribute their daily activities to the correct client without manual data entry, an enhancement using technology not in common use within the legal space but in regular use by lawyers who wear smart watches. **Disruptive Innovation.**

Innovation Spectrum to Market Understanding Spectrum

INNOVATION SPECTRUM

	REACTIVE	RESPONSIVE	INCENTIVE	DISRUPTIVE
DEFINITION	Reactive Response	Additive Upgrade	Portfolio Expansion	Opportunistic Diversion
EXAMPLES	Improved Interface, Bug Fixes	New Features, New Integrations	New Ecosystem, New Pricing Model	New Solutions, New Problems Solved
PRINCIPAL DRIVERS	Customer Directive	Market Expectations	Organization Strategy, Market Anticipation	Trends, Non-Linear Synthesis
BENEFITS	Customer Intimacy	Share of Wallet	Market Expansion	Industry Leadership, Singularity
CHALLENGES	Prioritization, Disruption	Proactive Outreach, Adoption	Focus, Awareness	Scale, Funding
RISK/REWARD	High/Low	Low/Medium	Medium/High	High/Very High

MARKET UNDERSTANDING SPECTRUM

	REACTIVE	RESPONSIVE	INCENTIVE	DISRUPTIVE
PROCESS	Listen-React	Listen to Solve	Analyze to Expand	Synthesize to Lead
OBJECTS OF LEARNING	Users - Workflow	Users - Function/Buyers, Competitors	Organization Strategy, Markets, Enablers	Organization Vision, Ecosystem, Markets
THEME	"Tell me what you want."	"What was you want?"	"What do we want next?"	"What do we dream of?"
SOURCES	Customer Service Calls, Sales Requests	Buyers, Regulatory Agencies, Advisory Boards, Win/Loss, Monitoring Negotiations	Market Leaders, Technology Advances, Workflow Mapping, Expert Updates, Trend Analysis	Cultural shifts, Ecosystem Changes, Societal Discussions, Demographic Analysis, Broad-Source Review
LEARNING TOOLS	N/A - passive	Proactive information gathering	Concept testing	Synergistic Thinking
WORKFLOW	Hear it - add to backlog			Multi-source synthesis & ideation

INNOVATION SPECTRUM

Aligns teams around what innovation means in your organization.

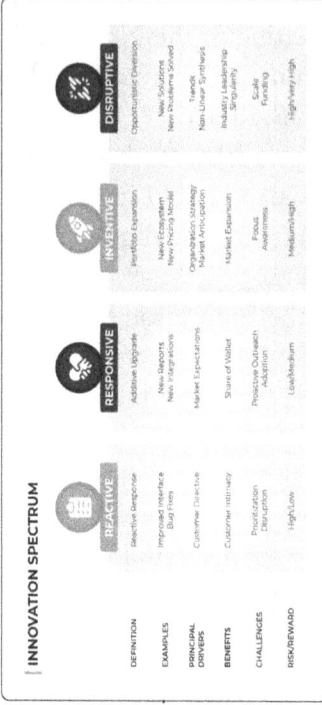

MARKET UNDERSTANDING SPECTRUM

Offers ways to find the Right Idea to innovate on purpose.

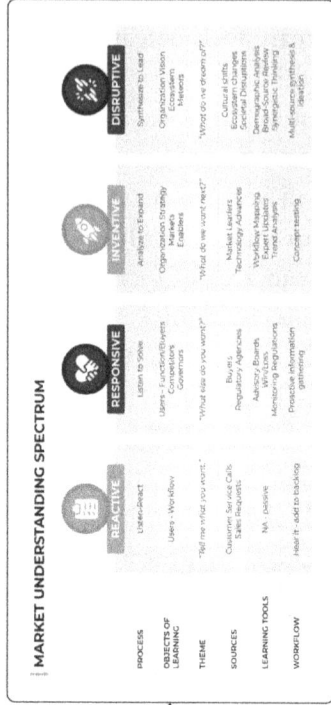

There are good reasons for organizations to focus on Reactive Innovation, just as there are for others to look for Disruptive Innovation opportunities. But you as the Market Strategist have to know which one it is, in order to innovate on purpose.

In other words, you need to find the *Right* Idea, not—necessarily—the Big Idea.

THE MARKET UNDERSTANDING SPECTRUM

Now that you have alignment on the Innovation Phase you should be focusing on, you can get busy brainstorming activities to get there. To bring the right ideas, you have to listen to your markets the right way.

Different levels of innovation require different levels of market knowledge, including different methods of market research, focused on different individuals and sources, and asking different questions. For example, Inventive and Disruptive Innovation are largely driven by your organizational strategy and where the market writ large is headed; Reactive and Responsive, more focused on what existing customers of existing products want now or soon.

Below is a companion tool to guide your market research planning, based on the phase of the Innovation Spectrum you want to play in. This tool is the **Market Understanding Spectrum**. The phases are the same as the Innovation Spectrum, but deliver guidance on how to think about performing your market research. It includes high-level direction on what process to follow, who and what to learn about, key themes, sources and learning tools as well as suggestions on the workflow to get results.

> **The Market Understanding Spectrum is a companion tool to the Innovation Spectrum and details how to listen to the market to get the right insights to deliver on each phase of the Spectrum. The phases on both Spectrums are the same.**

Budgets, resources, scope, and risk all escalate the farther to the right you go on the Understanding Spectrum, as do timelines for opportunity and ROI.

Phases of the Spectrums

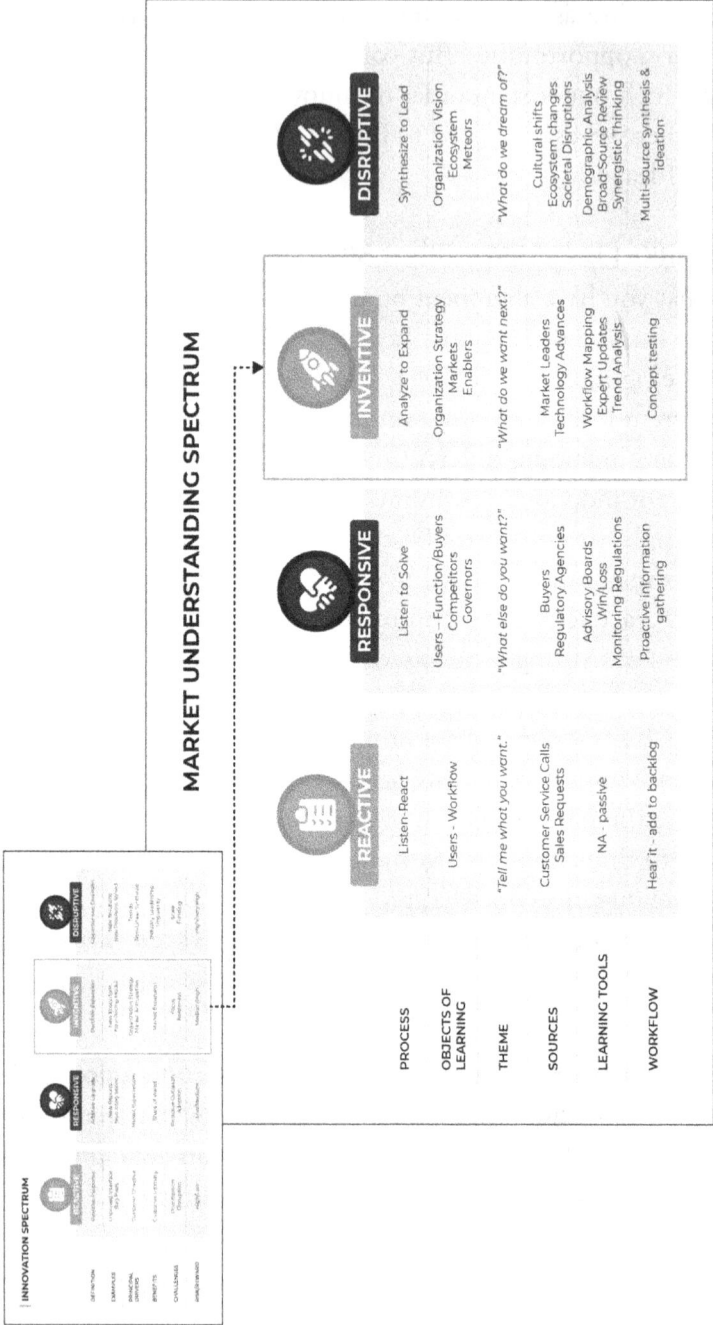

MARKET UNDERSTANDING SPECTRUM

	REACTIVE	RESPONSIVE	INVENTIVE	DISRUPTIVE
PROCESS	Listen–React	Listen to Solve	Analyze to Expand	Synthesize to Lead
OBJECTS OF LEARNING	Users – Workflow	Users – Function/Buyers Competitors Governors	Organization Strategy Markets Enablers	Organization Vision Ecosystem Meteors
THEME	"Tell me what you want."	"What else do you want?"	"What do we want next?"	"What do we dream of?"
SOURCES	Customer Service Calls Sales Requests	Buyers Regulatory Agencies	Market Leaders Technology Advances	Cultural shifts Ecosystem changes Societal Disruptions
LEARNING TOOLS	NA – passive	Advisory Boards Win/Loss Monitoring Regulations	Workflow Mapping Expert Updates Trend Analysis	Demographic Analysis Broad-Source Review Synergistic Thinking
WORKFLOW	Hear it – add to backlog	Proactive information gathering	Concept testing	Multi-source synthesis & ideation

SCALE OF INNOVATION: HOW MUCH RISK
SHOULD YOU TAKE TO INNOVATE

Now that we've translated the right phase of the Innovation Spectrum into some actionable guidelines for market research using the Understanding Spectrum, it's a good time to talk about risk. In the case of innovation, risk is found in the amount of market knowledge you gather before and as you build and take solutions to market. Gather too much information without acting, and you'll fall into analysis paralysis. Jump on any comment from a user, and you'll spend too much time building things (almost) nobody wants.

What's the *right* level of market knowledge? It depends on your organization's tolerance for risk. What you need to quantify is the level of knowledge necessary to justify market action in *your* company.

At the highest level, leadership and Market Strategists need to agree that certain knowledge becomes actionable when it's been documented to a predefined level of satisfaction. What is that? Only you and your leadership can say. But defining the thresholds gives not only you but the entire organization a shared understanding of "what we know," and what the level of risk is in acting on information.

To accomplish this alignment, I recommend creating a **Confidence Language**, which we'll define and complete in Chapter 8. I'll use an example with three knowledge tiers: Believed, Perceived, and Recognized:

Believed. This is knowledge that you believe is true but don't have any market research or documented experiences to back it up. A weakness in Believed knowledge is that, without documentation, it's difficult to get funding or rally a team around it. Believed knowledge represents the highest level of risk.

Believed, Perceived, and Recognized

BELIEVED, PERCEIVED, AND RECOGNIZED

BELIEVED (RED) — "We believe this is true but have little or no data to support that belief."

PERCEIVED (YELLOW) — "We have some data that tells us this is true, but not as much as we want."

RECOGNIZED (GREEN) — "This meets our data threshold of truth – it is a recognized truth."

Perceived knowledge is when there is **some** documented evidence that the perspective exists in the market, but the backup data is so slight as to make it unactionable. It's recommended that Perceived knowledge be validated with experiments or follow-up research. Acting on Perceived knowledge is less risky than on Believed but is still not the sweet spot for most companies.

Recognized knowledge has reached a threshold set by your organization to indicate that the knowledge is *actionable*—you can and should use it to shape market responses. Recognized knowledge has been quantified, or heard from specific sources that are considered reliable, or from enough sources to establish a representative sample of the market.

Leveraging a common language for defining what you know about your markets helps telegraph to leadership, teams, and new hires your relative certainty that something said about a market is true. Actions can be taken based on this common understanding and level of risk. While it's important to mention the need for such alignment here, I'll detail how to create a Confidence Language later in the book.

What else do you need to align with leadership on to create a contributing product strategy?

SCALE OF INNOVATION: AMOUNT OF COMMITTED RESOURCES

Gaining clarity on where the company's going and how you fit in, what a Big Idea looks like, and how much risk leadership is comfortable taking when they invest in going to market already have you far ahead of most Market Strategists. The next step is understanding what resources are available to get there.

Notice I didn't say the next step was "getting a budget to achieve the dream." If you have influence over budget allocation, wonderful—there are lots of tools to help you build out a budget and get approval. But many of you aren't that lucky. What can you do?

Chances are, your leaders already have budgets for market research, product development, and marketing in place, and you'll need to work within them, at least for now. That doesn't mean, however, that you shouldn't have a conversation about how resources should be allocated. You should, and here are some of the areas you should get clarity on:

- *Your time.* As the Market Strategist, it should be assumed that your primary role is searching for market-driven innovation that contributes to organizational success. That said, if you perform multiple roles for your company, you may need to explicitly agree on how much time you'll spend on each. This is highly advisable to manage expectations on how much market knowledge can be gathered, analyzed, and communicated.

- *Others' time.* How much of your data science or development team's time and resources can you expect to assist in finding innovation opportunities? How much can you expect from sales and customer service? There are many channels inside the organization that you need to rely on for information; the more they can help, the better and faster innovation will be achieved.

- *Research budget.* Do you have a budget for outside research? Travel? Professionally moderated focus groups? Do you have authority to spend it, or do you need approval from someone else? Set expectations up front that the farther to the right you go on the Innovation Spectrum, the higher the reward—but also the higher the budgets and resources to get to an opportunity. Listening to Disrupt will be significantly more expensive, time consuming, and risky than ongoing focus groups that help you improve an existing product.

- *Funding for test and build.* This may seem like putting the cart before the horse, but it's worth the time to have a brief conversation with leadership about how they would enable the pursuit of opportunities you discover. If you find a disruptive opportunity, who will build, market, and offer the beta offering? Does the organization have funds earmarked for such initiatives, or would you have to raise funding through venture capital or loans? If you can't spend $5M to launch and scale a new product line, it's best to know that up front. Sometimes the rhetoric of Big Ideas falls to the wayside after a discussion of fiscal realities but only if you *have* the conversation.

Once you reach alignment in these areas, remember to report on your adherence to them. This allows you to not only manage your own time but remind your leadership and the rest of the organization what you're *supposed* to be spending time on and why.

Your contributing vision and strategy statement is a foundational element to innovating in ways that are not only good for the market but good for your company. But there are a few additional expectations before you complete it, including:

- *Revenue expectations.* In hindsight, I realize how lucky I was to have revenue and profit goals attached to my product responsibilities. It was added pressure, but it gave me a window into how the leadership team thought, and aligned me with corporate expectations. I recommend you get this understanding too, whether you're evaluated on the financial performance of your offerings or not. How much revenue is "big"? I've worked for organizations that were interested in ideas with $5M in projected sales by year 3, and others where the CEO specifically told us, "Don't bring me anything less than a $50M opportunity." You may also be the Market Strategist for a product whose strategic importance to the company lies in maintaining the status quo: retaining the customers (and revenue) you already have.

- *Non-revenue expectations.* Maybe revenue isn't as important as fending off the competition or opening a new market. Or, for those of you looking to be disruptive innovators, perhaps revenue isn't the only reason for these radical new directions. Maybe, for example, it's to reinforce your brand as cutting-edge and experimental. For mission-driven organizations such as charities, you may be thinking in terms of impact or value, not revenue. Be sure to gain an understanding of these non-revenue benefits too.

- *Timeline.* When do you expect an innovation to deliver revenue? Profit? Breakeven? Whatever the strategic goals, discuss what milestones leadership will be using to update the company owners. In most of the organizations I worked with, checkpoints with the board or owners were quarterly, for example.

BONUS ALIGNMENT: UNDERSTAND *WHY*

I have one final thought before you craft your contributing vision and strategy: This is a great time to understand more about the *way* your leadership team thinks—to get inside their heads. Depending on the time you have left in your meeting and the cultural transparency in your organization, you could ask a few of the following questions:

- Why are we focusing here and not elsewhere? Why were other options rejected?

- What do you see as the biggest internal hurdles to success? The biggest external risks?
- How is the leadership team communicating with each other and the teams to reinforce the direction? How will leadership communicate progress with ownership?

The purpose of this line of questioning isn't to push back on decisions that have already been made but to give you insight into why other directions were rejected.

YOUR RESPONSE: A CONTRIBUTING STRATEGY

By having this discussion, you're better prepared to innovate on purpose than 99% of Market Strategists! But to move from alignment to leadership approval and cross-functional team enablement on what should happen day-to-day, you need to articulate your contributing strategy. Likely, this will take many forms: a long form being a business or strategic plan; the shorter form a contributing strategy statement in the same format as your corporate vision and strategy.

For our purposes, I'm going to offer some guidelines that split the difference. Let's use our LegalcoEG example from earlier in this chapter. For reference, here's the same *corporate* vision and strategy statement we used for LegalcoEG :

*As the premier provider of time-and-billing solutions for South American law firms, our **vision** is to help law firms everywhere increase their effectiveness, efficiency, and profitability through better workflow enablement tools as the go-to company for the most sophisticated solutions for running a law firm.*

*Our **strategy** to achieve this includes expanding our product line to offer end-to-end workflow solutions, first for law firm finance departments and then for all aspects of running a law firm. We'll enter the Canadian and US markets while assessing next steps for further globalization. We'll expand our solutions beyond software to integration, implementation, best-practices consulting, and other offerings to become the trusted back-office partner for global law firms.*

As in the earlier use of this example, you are the Market Strategist responsible for time-and-billing products. Using all the elements discussed so far, here's one way you might articulate your contributing strategy:

- **Who you are.** Start by stating what products or markets you're responsible for so everyone understands what point of view you're coming from. Example: *"We're the group in charge of the time-and-billing software product."*

- **How you contribute.** Express your responsibility for fulfilling the corporate vision. Example: *"In the future, we'll continue to focus on the current software offering."*

- **From this to that.** In larger organizations, many folks won't know what you do today, much less what you're supposed to become tomorrow. Give this context here. Example: *"We're charged with moving from solving the time-and-billing challenges of South American law firms to also enabling the finance departments of law firms in Canada and the US, and ensuring the software interoperates with the other workflow-enablement solutions we're building for law firms."*

- **How you'll get there.** This should be three to five points from your strategy, highlighting the most significant efforts you and the team will undertake to achieve this vision. Example: *"To achieve this, we'll test our existing product in the Canadian market to establish a baseline, then use what we learn to modify and built new features to compete in that market. We'll meet with our fellow LegalcoEG Market Strategists and managing partners of law firms in our key markets to understand the requirements of interoperability."*

- **More of/less of.** Point out specifically how some activities you used to do will be slowed or stopped in order to focus on the new. This is important to enable teams to prioritize day-to-day activities. Example: *"To achieve this vision, we'll be spending more time on market research and testing with prospective Canadian product users and buyers, and less time building features for our existing South American customers. To compensate for the reduced focus on current customers, we're going to home in on understanding and executing on the priorities of our biggest current customers."*

- **Time on BAU versus vision.** I recommend a percentage, to give direction without being too specific. Important to enable everyone on your team to make their own time-management decisions. Example: *"For the next 12 months, our goal is to spend 40% of our time on the new initiatives and 60% on the roadmap currently being executed for our South American customers; we'll be adjusting the ETA on those milestones as a result of this resource reallocation."*

- **Outcomes and measures of success.** Reinforce the endgame of the vision and how you contribute to it. Example: *"Our efforts should result in opening the Canadian market with our first sale in Q2 and direct revenue of $30M by 20XX. We'll also enable the success of our new product offerings with seamless integration and interoperability."*

Here's an example pulled together as a contributing strategy statement by the Market Strategist for the time-and-billing team at LegalcoEG:

As the team in charge of our time-and-billing software product, we're responsible for maintaining the success of the current offering with existing customers and expanding success into law firms in Canada. While our other teams will be focused on finding new problems to solve in these markets, we'll ensure that our software interoperates with the new products as part of a seamless, end-to-end solution.

To meet market-entry goals, we'll test our existing product in Canada to establish a baseline while researching the market in depth.

We'll look for Inventive Innovation opportunities to pivot around existing-but-outdated competition. To achieve interoperability, we'll meet with fellow LegalcoEG Market Strategists and managing partners of law firms to understand requirements.

Because our resources remain the same as last year, we'll spend more time on market research and testing with prospective Canadian users and buyers and less time building features for existing South American customers; the goal is to spend 40% of our time on the Canada expansion, 10% on interoperability, and 50% on the roadmap for South American customers.

This is a 12-month plan. We expect to open the Canadian market with our first sale in Q2 and revenue of $30M by 20XX, while maintaining South American renewals. We'll also enable the success of our new product offerings by meeting integration and interoperability needs.

And there you have it—a contributing strategy. Whether you build this as a draft and take it to your leadership for input or work on it together, ask these questions and build the response to lay the foundation for successful innovation.

With this knowledge, you can make a plan, focus a team, and innovate on purpose.

Now that you understand your organization's strategy and have articulated how you'll contribute to its achievement, it's time to assemble your team.

ACTION PLAN FOR CHAPTER 4

- Review your organization's corporate vision and strategy, if there is one, and use it to draft an answer to the four "who we are/who we want to be" questions for your entire company. No formal vision and strategy? Then answer the questions to the best of your ability without it.

- Using the work above, carve out what you believe to be your responsibilities from the overall corporate. As you document your strategy to get from who you are today to who you want to be, give some examples of what you'll need to do more of and less of in order to get there.

- Map your assumptions about innovation phase from the Innovation Spectrum in Chapter 2 to the Understanding Spectrum in this chapter. Brainstorm a few ideas of what this means to you in terms of:
  - Market research priorities;
  - Current market research and product roadmap plans; and
  - Team assumptions about ideation.

- What is the appetite for risk in your area of responsibility? Make a note of what you assume it to be.

- From the above brainstorming, craft a first draft of your contributing strategy statement.

- Meet with your leader to discuss and edit.

- During this meeting, align with your leader on resources available, timeline, and goals. Add these to your final contributing strategy.

- Share the contributing strategy with your cross-functional team, and refer to it to provide guidance on priorities and tactics.

Chapter 5
ASSIGN
OWNERSHIP

Innovating on purpose is a nuanced, sophisticated activity, but the roles and workflow to do so should be simple and straightforward. There are several self-inflicted roadblocks to purposeful innovation, and a big one is lack of ownership.

WHAT IS OWNERSHIP?

Ironically, most dictionary definitions of "ownership" include the word "possess." The implications of *possession* don't really fit in our workflow. Perhaps that's because ownership, in the world of business, is defined within teams. Possession, to me, is all about **having**. To own innovation—you gotta **give**.

To innovate, you have a role to play that depends on and serves other roles. All team members, in turn, deliver on a common—organizational—goal. In the world of market-driven innovation, you have to deliver what the market you've chosen to serve wants and needs. Ownership isn't about possession: it's about responsibility.

As the owner of an activity, what are you *responsible* for? At the highest level, you should make sure that you:

- Understand what you're being asked to do and how it fits in with the big picture;
- Acquire the skills to perform your role and complete your deliverables;

- Perform your role to the utmost of your ability;
- Listen to others who may contribute to the excellence of your deliverables;
- Assist others as required to excel in *their* deliverables;
- Follow strategic guidelines by focusing on your responsibilities, while being flexible enough to help in unexpected situations;
- Communicate the thinking (and data) behind decisions you make; and
- Know what outcomes you're expected to achieve.

With that explained, we can move on to answer the questions at hand: What needs to be done to innovate on purpose, and who should do it?

WHAT TO DO TO INNOVATE ON PURPOSE

What needs to be done to innovate on purpose? The overarching goal of the activities below is to bring together business strategy and market insights so you can build products and services at the intersection of what your market needs and who you want to be. Those activities include:

- *Understand the corporate vision and strategy and respond with a contributing strategy.* Both of these were defined in the last chapter; here in Chapter 5 I'll reinforce who owns what in these steps.

- *Create an innovation map.* This is the transition from laying out a contributing strategy to identifying what market research needs to be done to achieve it. Someone needs to lay out the plan to learn about the market, including assembling a prioritized set of market research activities and creating a budget, schedule, and backlog to deliver opportunities for market-driven innovation.

- *Create a data structure and choose tools to analyze the market.* Market data is only good if it's used, and to use it someone needs to organize it and put it somewhere where it can be accessed and analyzed to make decisions. This requires standardizing on language and structure of the data you gather and choosing what tools will be used to aggregate, analyze, and prioritize.

- *Gather information.* The good news is there are myriad channels from which excellent, actionable market information will flow to inspire innovation. The bad news is, most organizations aren't taking advantage of them because no one is acting as the champion for systematically gathering or consistently aggregating the bits of insight as they come in. We'll assign someone to do that.

- *Manage aggregation.* Whether you are able to afford one of the elegant software tools available to aggregate data, facilitate analysis, and visualize opportunities or are making do with spreadsheets and a basic file-hosting app, someone needs to make sure all the data trains arrive at the station. Even with the most powerful tools available, data aggregation may include putting hands on a keyboard to transcribe insights or clicking through links to find and compile original data sources.

- *Perform analyses.* This activity involves the technical, mathematical, programming skills of data mining—the search for predictive correlation, patterns, and trends. The individuals who own this work are likely **not** the ones who will **use** the knowledge mined, but they're critical to getting the most value out of what's there.

- *Hypothesize.* The person who owns this activity is responsible for turning market data into hypotheses for further refinement, creating actionable statements of knowledge, and acting as the voice of the market in various ideation and product/response processes.

- *Test findings.* Continuous questioning of processes, definitions, and conclusions drawn from market data will make the process stronger and build cross-functional trust, so having someone formally assigned to this task telegraphs into your organization that you take it seriously. Note that this is different from testing products or messaging in the market—it's testing what you think you know about the market so you can respond better.

- *Communicate.* Everyone, from the legal team to the developers to your billing department, does their job better when they understand the market. Information needs to be centralized to be understood, but it needs to be communicated to be leveraged. Who will deliver on that promise?

WHO DOES WHAT

If you've got an organizational structure in place that's working for you, great—keep using it. If ownership is already clear and these deliverables are getting done, you're good. If not: here are my recommendations for who should own what:

THE CEO/EXECUTIVE LEADERSHIP TEAM:

As discussed in Chapter 4, your senior-most executive defines the corporate vision and strategy for the entire organization: the "who you want to be" element of innovating on purpose. They structure the teams, set financial parameters, and define what success looks like. As teams respond to the organizational strategy, executive leadership acts both as evangelist and enforcer of that direction.

THE INNOVATION TEAM:

The innovation team responds to the organization's vision and strategy. Focused by direction given by leadership, this team plans, delivers, analyzes, and communicates the market knowledge to achieve that vision. The team includes:

Market Strategist:

As stated earlier, the Market Strategist responds to the corporate vision and strategy with a contributing strategy, and leads the work to achieve by delivering innovation opportunities to the organization and, at a high level, shepherding them through the build and commercialization process. This role is responsible for looking to the future: crafting a holistic view of the market but also how changing norms in society, demographics, and general patterns of buying and using products come together to provide opportunities to their organization. Having a Market Strategist is necessary if organizational objectives include Inventive or Disruptive Innovation and is strongly recommended for responsive innovators as well. The Strategist creates the plan and leads the gathering, aggregation, and analysis of market data and other knowledge needed to innovate to achieve business goals, prioritizing both the research to be done and the opportunities discovered. The Strategist also communicates opportunities to the organization in the form of actionable insights so they can use their expertise to design, build, price, sell, and support solutions.

In short, the Market Strategist is responsible for driving the intentional innovation workflow laid out in this book.

Product Management and Product Marketing

These roles work closely with the Market Strategist—possibly *for* that person—and are responsible for the short-term opportunities for Reactive and Responsive Innovation. They are charged with being the experts on product users and buyers, respectively, and working with design, development, and marketing communications to get product improvements into the hands of customers. These team members will work closely with the Market Strategist in all aspects of the innovation workflow.

Data Engineer/Data Scientist/ Information Services Team:

Based on the requirements (and budget) of the Market Strategist, these individuals establish the appropriate taxonomies, standards, platforms, and tools to enable market research. It could be an integration of functional workflow-enablement systems such as Salesforce and Zendesk into an aggregation and insights tool like Aha, Confluence, or productboard, or something as simple as an Excel spreadsheet with a few pivot tables with links to materials stored in Dropbox.

Data Analysts/Market Researchers:

These individuals are trained to mine data. While the Market Strategist directs the goals and objects of research, the researchers are the experts on how to get the insight that's needed. Researchers also help by crafting survey and focus group questions that lead to actionable knowledge and trusted results.

Data Wranglers/Data Entry/Data Aggregator:

Data entry—what is this, the 1990s? Even though much of the data you need is automatically gathered or generated through systems like those mentioned earlier, some will not be. Sales calls, focus groups, articles may need to be scanned or entered manually to ensure truly robust insights. Having owners of this work—literally hands on keyboards—expands the variety of sources available to you and enables you to be more flexible in how you ask others to provide the knowledge they gather. Those with higher-level skill sets can also help by tweaking matching rules, cleaning up errors, and other data organizing tasks.

THE CONTRIBUTING TEAMS:

Contributing teams are functional experts within the company and play their own role in delivering purposeful innovation. Whether sales, designers, content writers, or members of the finance team, they have a full-time job to do, based on their specific expertise, that isn't about creating something to sell in the market.

That said, these functional experts glean information about changes in the market, or the environment the market exists in, that could be valuable to discovering opportunity for innovation. We want to capture all these insights.

To make that happen, all team members should be held responsible for contributing what they hear and observe to the aggregated market knowledge. But it's important for the Market Strategist and their team to make this as easy as possible—after all, these folks have a primary job to do, and asking them to provide market insights is acceptable but should be structured to remove as much of the burden from this responsibility as possible.

Most organizations at least try to aggregate market knowledge from sales, customer service, and the results of email campaigns and other marketing efforts, but there are other rich knowledge sources you can learn from, including:

- *Design.* Depending on your industry and organizational structure, your user experience or product design group may be the ones closest to the current users of your products, and the people who best understand the logistical and physical challenges of those users. They are a great source of knowledge for Reactive and Responsive opportunities.

  That said, the more strategic knowledge to mine from your design team is what they learn *as part of their own ongoing functional training.* They are experts in emerging best practices to solve the product design needs of tomorrow. In other words, they have knowledge on the "tool kit" your company has at its disposal to pursue the opportunities you find in the market.

- **Legal.** Often, your corporate counsel is involved at the last phases of deal negotiation, working over specific points in the contract. What portions of the contract are most often disputed or redlined during negotiations? Who besides the customer's legal counsel gets involved with contract negotiations? Your in-house counsel also serves as your expert on all things legal. It may or may not be feasible to have your corporate counsel monitor legislation and litigation ongoing in your markets, but if you **can** leverage them to do that work, you will get better results than trying to do it yourself. You should also tap them for periodic updates on the regulations you and your markets are governed by.

- **Implementation teams.** What are the current systems and platforms our customers expect you to integrate with, and when do they plan to upgrade? What Internet browsers (and versions of those browsers) need to be supported? Are they thinking of changing the platforms or services they're using today? What about the actual implementation process. Is the amount of time it takes acceptable? How about the expense and indirect expense of their own resources being tied up in the process? Would they rather spend more and get more implementation help or just the opposite?

- **Marketing communications.** All manner of buyer journey data, including email open rates and click-throughs on A/B messaging tests, survey results, landing page performance, and webinar attendance, can help you understand how your market becomes aware of you and buys your products. Marcomm should also be able to give you insights on what types of marketing tools, channels, and methods are emerging and work with you to determine which ones to test. And don't forget that these folks are design and communication experts—not just for external audiences but for internal too. In large organizations there is often a small group specifically responsible for communicating **inside** the organization. This will be a valuable team for you as well, as you endeavor to distribute this market knowledge throughout your company. That team can not only help you create those materials but give you insights on the internal "market."

- *Accounts receivable.* Why are customers refusing to pay their bills? Is it a complaint about the product, or are they going out of business/not doing well financially themselves? Are they complaining about the price they've been charged? Each of these indicates a very different threat/opportunity to your organization, and each should be handled with a different, innovative response.

- *IT/R&D/the lab.* The knowledge you can gather here is primarily about the tool kit—what technology is available to pursue opportunities? You don't need to know everything about the tech available to create products to serve your markets (that's their job), but it is wise to have an idea of how that world is changing, to open your own perspective on what opportunities are within your reach to pursue when it comes time to prioritize and ideate. It's important to create an expectation that your IT team will not only deliver a solution to meet the market's need but do so in a way that fits the market's technical expectations *and* limitations.

GAIN CLARITY WITH THE INNOVATION ALIGNMENT TOOL

So far, I've offered some examples of team roles and the activities typically performed by them. But when was any organization typical?

If you aren't sure who does what when it comes to innovating on purpose, use the Innovation Alignment Tool shown below. This is a simplified RACI exercise that helps you align on who does what in the innovation workflow, whatever the size and structure of your organization. Get the individuals who *could* be owners of each activity together and determine owners and assists.

Innovation Alignment Tool

INNOVATION ALIGNMENT TOOL

	DEFINE ORGANIZATION STRATEGY	CREATE CONTRIBUTING STRATEGY & INNOVATION MAP	DEFINE DATA STRUCTURE, CHOOSE ANALYSIS TOOLS	GATHER DATA	AGGREGATE DATA	ANALYZE FOR INSIGHTS & OPPORTUNITIES	TEST FINDINGS PERFORM RESEARCH	COMMUNICATE INSIGHTS, ACT AS MARKET EXPERT
OUTCOME	Market Strategist is focused on what drives vision & strategy	Innovation opportunities uncovered deliver on organizational strategy	Market data is easily accessed & analyzed, drives successful innovation	All pertinent data sources are gathered; have a full picture of innovation opportunities	All available data is accessible to drive successful innovation	Find market needs that drive organizational success	Confidence that we've minimized risk and maximized opportunity for market impact	All departments have the market knowledge needed to deliver innovation
OWNER	CEO	CEO	Data Scientist	Market Strategist	Data Wrangler	Market Strategist	Data Analyst	Market Strategist
ASSIST	Org Owners, Market Strategist, Functional Leads	Market Strategist	IT, Data Engineer, Data Scientist	Cross-Functional Teams, Data Scientists & Wranglers, Data Entry	Data Entry, Data Analyst	Data Analyst	Market Strategist	CEO Functional Leads, Internal Comms
TOOLS	Owner goals, results of market research	Listening Map, Workflow	Taxonomies, Python, Tableau, productboard	Salesforce, Teams, email, vm, texts, hard copy	Aggregation Tools chosen by Data Scientist	Aggregation Tools chosen by Data Scientist	Focus groups, surveys, A/B testing, primary source review, analyst reports	Tableau, meetings, videos, Teams Channel

Also ensure that you've aligned on what outcomes each activity will strive for and the tools you'll use to get there. As you start out, you may have very rudimentary tools for gathering and analyzing data—that's okay! Don't wait for budget approval for the shiny, sophisticated system. Using the tools you have available is a great way to prove concept, making it easier to get the system you want as you show results.

Once the team is in place and ownership of activities defined, you can turn your attention to building out the framework to organize your market knowledge.

ACTION PLAN FOR CHAPTER 5

- Are roles and responsibilities already clear? Make sure your cross-functional team has that information.

- If not, meet with your cross-functional team and, using the Innovation Alignment Tool, complete what you all believe is the right role/responsibility distribution.

- If there are conflicts, propose switching off ownership every six months to determine where the ownership fits best. If the team can't agree on this, note that the responsibility is currently unowned.

- Have each member of the team take the draft to their leader for input and approval.

- Finalize and publish your team's Innovation Alignment Tool. Revisit as needed: I recommend doing so with any reorganization or major shift in product/market responsibilities.

Chapter 6
DEFINE A KNOWLEDGE STRUCTURE

Determine Roles & Goals

- Align with Leadership
- Assign Ownership
- Define Knowledge Structure

Finding innovation opportunities isn't as easy as asking users what they want you to build. But it's a waste of time and resources to wander through the world hoping an innovative idea appears in the chaos. Charlie Munger, vice-chair of Berkshire Hathaway, said, *"Opportunity comes to the prepared mind."* I say a prepared mind is one that focuses on the parts of the market that are most likely to deliver opportunities that help achieve the corporate vision. To enable that focus, you have to not only intentionally choose where and how you research the market but how you organize the data you find there.

It's time for the last step in defining roles and goals: defining a structure to organize what you know about your markets so you can use it.

A caveat: I'm not a taxonomist, a data scientist, or a data architect. If you have those resources available, you should leverage their expertise to create your data structure and decide how to gather, aggregate, and access your data. You may even already have a business information system or set of tools to use.

My intention is not for you to take on that responsibility. But as the Market Strategist, you're the party most invested in ***using*** well-organized data to do their job. Therefore, you may not be the individual who creates the data structure, evaluates the aggregation and visualization tools, or crunches the data, but you *are* the one whose requirements must be met by doing all of the above. And if it's your requirements that must be met, you have to be responsible for—and prepared to—set them.

I want you to have a way of thinking about how to organize your current and future market knowledge so you can use what you learn to innovate on purpose. You can keep these ideas in mind when you evaluate tools, or work with your data scientists, or cobble a system together yourself. For the Market Strategist, it's critical to use *some sort* of data structure that enables ongoing data aggregation, visualization, and analysis. At the end of this chapter, you'll have all of the above.

TYING INNOVATION ASPIRATIONS TO A PLAN FOR MARKET UNDERSTANDING

In Chapter 4, we discussed the need to align with leadership on the corporate vision: specifically, which markets, market problems, type of product, and market position you want to focus on and achieve. With that alignment to guide you, you can plan your market research to proactively seek out innovation opportunities that will deliver the type of success–the right ideas–to realize it.

Too often, we waste the market knowledge we gather, because we don't organize it to maximize its use. Think about it: is the cost/benefit of your efforts to understand your market paying off? All those surveys, focus groups, data analysis—most of us still feel like we have more questions than answers and no real idea of how to grow in the present, much less lead in the future. When market research is not explicitly done to deliver on strategic goals, the waste is significantly worse.

These are self-inflicted roadblocks. Most companies are so busy *gathering* market information, they don't *use* it. They're so busy putting together surveys and focus groups, they forget to make a *plan.* And like assembling a team to innovate on purpose, assembling market data to enable achievement of a corporate vision and a specific type of market success is a sophisticated effort but a simple workflow.

To start: get organized!

Following is a structure—not *the* structure—you can use to organize what you know about your market so it's useful to your organization. I suggest grouping knowledge around objects in the market, including insights into attributes of those markets that may be opportunities for you, gleaned from reliable sources of information.

OBJECTS, ATTRIBUTES, AND SOURCES

Whatever organizing structure you choose should allow you to discover knowledge three ways:

- Access data by specific personas and markets—the targets of or influences over your organizations vision of who you want to serve—both at aggregate and discrete levels. These are the **objects** of your investigation.

- Aggregate data that gives you specific types of knowledge about these targets. This could include the needs of a persona, the strength of a market, or the regulatory constraints of your possible response. These are the **attributes** of the objects you investigate.

- Know and understand the relative value of each of the **sources** of data you gather. You'll want to know the quality, currency, and reliability of any data source to determine how much weight to give it in your considerations. Data points collected over time can also give you insight into trends.

Below is a visualization of objects, attributes, and sources, with a definition of each:

**OBJECTS, ATTRIBUTES, AND SOURCES:
AN INFORMATION STRUCTURE**

OBJECTS	ATTRIBUTES	SOURCES
TARGETS of learning. You learn about them to SERVE, SELL or otherwise IMPACT them. Objects may also impact YOU.	ACTIVITIES OR CHARACTERISTICS of an OBJECT. Attributes are what you learn ABOUT.	AUTHORITY from which you GATHER INFORMATION about attributes. A source MAY OR MAY NOT be an object.

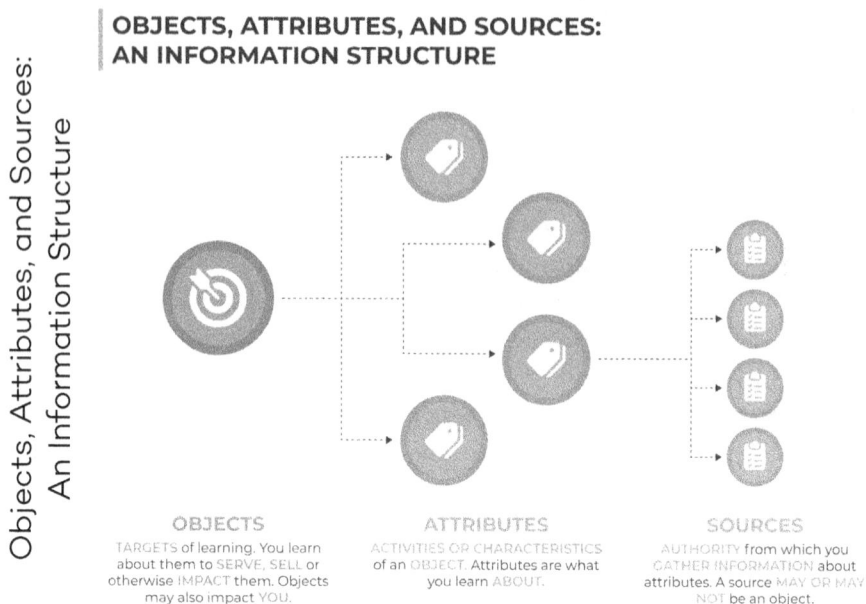

Let's explore objects, attributes, and sources in more detail.

OBJECTS: THE SUBJECTS OF YOUR INNOVATION RESEARCH

Objects are the targets of your investigation, and you learn about them to serve, sell to, respond to, or otherwise interact with them. Something is an object if you learn about it for its own sake, like product users, buyers, and competitors. And of course, your own existing products would be objects.

Most of your already have great (if poorly organized) information on those objects and, if you're tasked with maintaining the status quo in a mature market with an existing product—Reactive Innovation— they may be all you need to know about.

To innovate farther up the Innovation Spectrum, however, you may need to go beyond your current product, buyers, and users. While not an exhaustive list, here are some objects you could consider learning more about in your Innovation journey. I'll begin with the most popular objects—products, personas, and competitors—and then move on to some you may not have considered.

- **Products.** Products are what you offer the market. Typically, organizations define products as whatever it is they sell to customers, whether that's bicycles, banjo lessons, or banking software. That's a good first step, but you should also consider other elements of your offering that your markets might judge and evaluate separately, such as training programs or implementation services.

- **Personas.** A persona represents a group of people in a market that have similar characteristics. I recommend creating a persona based on their title or key demographic characteristic, for example "Ruth the sales rep" or "Ruth the mother of two." Here's why:

   As noted with products, personas can be nuanced, offering many types of opportunity, and requiring many types of messaging. To be valuable beyond a Reactive Innovation goal, you need to capture this nuance, which means organizing information around the varying ways you may interact with a persona/object.

   I recommend a secondary level of persona detail to add value and nuance to your investigation in two ways: persona functions and persona roles. We'll return to these in the upcoming discussion of attributes.

- **Competitors.** In addition to specific companies, competitors are any alternatives your personas consider instead of you. Some of you are lucky—you don't have any competition! But most markets include at least one alternative. Be sure to include indirect competitors such as "do nothing," "use a makeshift or indirect alternative," and "build it myself" in addition to more direct competition, if your market considers these options.

   For example: if your product is a lawnmower, your buyer may consider other lawnmowers (direct competition), hiring a lawn service (indirect competition), buying a goat to eat the grass (makeshift), building their own lawnmower (DIY), or just letting their grass grow (do nothing). To the degree these are options normally considered by your buyers, you should know what's going on with them.

Most organizations gather knowledge on these three objects. But what are some other objects you should consider keeping tabs on if you aspire to more than Reactive Innovation?

- **Opportunities.** Opportunities may become products in the future but, until then, you have to keep track of them as they emerge. Opportunities can come from the market needs you uncover, trends in the market, new laws or regulations, or your own strategic direction. Once you can articulate a strategic opportunity you can aggregate ongoing knowledge about it to help decide how and when to pursue it. Later we'll articulate opportunity statements, but in the meantime, think of these as "products that haven't happened yet."

- **Corporate vision & strategy.** Chapter 4 was about aligning with leadership to ensure your innovation efforts were focused on areas where leadership wanted to succeed. This is a reminder to formalize this alignment by treating your executive team and organizational strategy as an object to understand and serve.

- **Your contributing strategy.** Chapter 5 also asked you to formalize your response to the corporate vision and strategy with your own contributing strategy. Because this is the knowledge that focuses and aligns your cross-functional team, treat it as an object too.

- **Markets.** Markets are groups of organizations or people that perform a common activity. Examples include medical device manufacturers and Canadian Little League teams. By designating markets as objects, you can investigate market health (is the market growing or shrinking), logistic needs (how does this market want to buy, receive, and use your kind of product), and technical stack (what technology might you have to integrate with?).

- **Influencers.** Influencers enter workflows as people whose opinion or expertise your buyers trust to help them decide which product to buy. They **advise** your personas but don't **control** them or their decisions. Examples include industry analysts, subordinates on a team, small children in a family. You'll want to know where influencers appear in a workflow, the type and extent of influence they have, and over whom.

- **Governors.** Unlike influencers, whose input may or may not be heeded, governors are people and organizations that establish rules, standards, or regulations you or your markets **must** comply with. Governors can set rules for markets, personas, functions, and roles. For example: in the United States, the Food and Drug Administration (FDA) regulates food labeling and nutrition information. If you wanted to launch a new type of frozen pizza, your packaging would have to meet FDA guidelines, or grocery stores wouldn't sell your pizza.

- **Gatekeepers.** A Gatekeeper sits between you and your users and buyers, either as an intentional part of the supply chain or as an organizational barrier to your market. For example, let's say your product is a vaccine—you sell to a wholesaler who, in turn, sells to the pharmacy that sells (and administers) it to the person getting vaccinated. Another example could be executive assistants or procurement managers you sit between you and your buyers. Gatekeepers have their own needs and priorities that may or may not be aligned with, aware of, or dependent on yours.

- **Workflows.** Workflows are a standard group of activities performed by one or many people to achieve a result. All products are chosen, purchased, implemented, used, and renewed via one or more workflows that are pretty much the same for everyone in the market. There are different ways people choose whose pizza to buy and what kind of pizza to order; there are options to **get** the pizza, and even more than one way to **eat** the pizza. When groups of customers follow a pattern to achieve a result, it's a workflow you should understand and support.

Like markets, workflows are more than the sum of their parts. Also like markets, they can expand, change, and become obsolete over time. This is why considering workflows as objects to learn about and respond to is a powerful innovation strategy for existing products and potential disruptors.

Carvana showed us that consumers will buy cars online. Esurance showed us that they'd buy insurance online, and Ally Bank that they were okay depositing their money in a bank that didn't have a physical place to keep it. Contracts are out, month-to-month subscriptions are in. COVID changed product distribution expectations and buying priorities, and then some of them changed back—or morphed into something entirely new. Listening for what's changing now, whether it's a behavior your product has previously needed to reflect or not, will help you understand how to incorporate these changes into your offering.

By considering a workflow an object, you can investigate the of actions taken by one or many personas, influencers, or gatekeepers within a market: how buyers, users, and others come together to get things done.

- **Enablers.** Enablers are tools available to solve market problems, whether you've invented them or not. Some examples include cloud computing capabilities, the Internet, and QR codes. You should watch the market in general for ways to do something better than you do it today, especially if there are known market problems that, to date, haven't been solved well due to lack of tools to do so.

- *Ecosystems.* Ecosystems are the geographic territory and sociopolitical structures you and your markets operate in. How is the world around you changing in a way that might provide an opportunity to innovate? A current example is the increasing weight of environmental responsibility in choosing suppliers and products. Depending on your market or persona, you may want to ensure that your products meet green environmental standards. Note that ecosystem changes aren't those mandated by law but efforts put forward to signal to the market that you have similar priorities.

The type and number of objects you name and gather information on will depend on many factors, including the phase of innovation you aspire to.

With objects identified, you can inventory what you want to know about each.

ATTRIBUTES: WHAT YOU WANT TO KNOW

Attributes are what you want to know about the objects you study. Attributes can be characteristics of the object, such as age and education or market size and average sales, but they could also be questions you have or hypotheses to validate.

The attributes you investigate for each object will be different depending on the object and your innovation goals, but a good way to brainstorm the ones you want to capture (and communicate them to others) is to create constellations of attributes. Here's an example:

The following constellation reflects what I want to know about my target market, law firms. I've created "law firm market" as an object, and surrounding it, attributes of the market I want to know about. The examples I'm giving include the size of the market and the rate at which it's growing, what legal or professional governance they're constrained by, and how a typical law firm is structured. I've also included "billing trends" as an attribute, which would be likely if I was a creator of time-and-billing software for law firms. The attributes you research should have some bearing on how you serve or interact with the object.

AN ATTRIBUTE CONSTELLATION FOR LegalcoEG'S MARKET: LAW FIRMS

Law Firm
Structure

Billing
Workflow

**OUR MARKET:
LAW FIRMS**

Market
Strength

How
They Buy

Billing
Governance

Legal Services
Trends

Attributes are, as a rule, very specific to your markets and your innovation goals, so I'm not going to provide an exhaustive list of ideas. However, I *am* going to offer one suggestion when it comes to attributes for the object or persona: that you always include "functions" and "roles" as attributes for each persona you create.

PERSONA FUNCTIONS

Persona *functions* can be defined as the primary, ongoing responsibilities a persona in an organization fulfills to add value.

Functions may be reflected in a persona's title but not always. Let's stay with our prior persona example of "Ruth." Here are two examples, one B2B and one B2C just for variety, of functions Ruth may perform:

- B2B: Ruth the sales rep. Her *function* is to sell her company's products and deliver a certain amount of revenue to the top line.
- B2C: Ruth the mother of two. Her *function* is to raise the children in collaboration with her partner, if applicable.

PERSONA ROLES

Persona *roles* are *sporadic* activities performed by a persona, specifically those having to do with *your* interaction with the persona: how you work with or encounter the persona.

Using the B2B example of Ruth the sales rep, Ruth periodically participates in buying products like ours by influencing her manager to choose the one she likes best. Kafir, her organization's purchasing director, owns the *function* of buying products.

It's likely that you only care about one or two of Ruth's functions or roles; depending on how expansive the Innovation Phase you're going after, you won't gather insight on every one of her roles and functions.

Now you have targets of your investigation listed as objects, and questions about them in the form of attributes. The last element you need is an idea on where to get the answers.

SOURCES: WHERE YOU GO TO GET ANSWERS

Sources are authorities from which you gather information about an object or attribute. A source can also be an object: you may interview a managing partner in a law firm about how they record and store hours billed to a particular client. It could also be a purpose-made source of information like a survey, focus group, customer advisory board, or tracking tools built into your apps and offerings.

Most organizations over-listen to product users and over-listen to existing customers, another of the self-inflicted roadblocks to innovating on purpose. Of course you need to fulfill the needs of these folks. But if you aspire to grow beyond Reactive Innovation, you **must** get perspectives beyond theirs.

Being conscious of who you're listening to and having a plan to listen that matches your innovation goals is critical. Expanding your sources is often a quick way to jumpstart your ability to invent and disrupt. The sources that will help you most will be unique to your and your market's industry, but here are some ideas beyond your go-to:

- *Associations.* Often associations have deep trend and demographic analyses on the functions performed by your users or the markets themselves.
- *Regulatory agencies.* You may be able to set up a proactive monitoring service to get regulatory changes delivered to your Inbox. Often these changes mean new market needs that require mandatory resolution.
- *Observers / 2nd-person sources.* The influence a persona has and their place in a workflow may not be what they think it is. Asking others who were in the room or part of the process their perspective on the other players could provide valuable insights.
- *Governments.* Depending on where your markets operate, there may be a treasure-trove of data available at no charge from the local or country governments.
- *3rd-person innovators.* New technologies, pricing models, marketing tools and techniques, design innovations, and selling opportunities are being invented every day. Don't stick at finding such inspiration from other industries, and watch the emergence of new tools that could enable you to solve market problems in a whole new way.

USING OBJECTS, ATTRIBUTES, AND SOURCES TO INNOVATE ON PURPOSE

Innovating on purpose means aligning what you know about your market with your aspirations to serve it. Listening to be a reactive supplier is very different than looking for opportunities to disrupt. As a rule, the more disruptive you aspire to be, the more objects and attributes you'll investigate and the more sources you'll monitor. Think of the list of objects, attributes, and sources you're creating as your **OAS inventory**.

> **Your OAS inventory is a list or visualization of all the objects, attributes, and sources (OAS) you need to understand or investigate to achieve your contributing strategy.**

To innovate on purpose, create—and maintain—an inventory of objects, attributes, and sources—a list of places to get the information you need to innovate for the markets and personas you want to serve. The simple step of creating this inventory focuses your thinking and allows you to prioritize scarce resources to gather knowledge you can use, even when you're not completely sure what you're looking for.

OBJECTS OF OPPORTUNITY RESEARCH
ALIGN WITH PHASE OF INNOVATION DESIRED

REACTIVE	RESPONSIVE	INVENTIVE	DISRUPTIVE
Reactive Response "Listen – React"	Additive Upgrade "Listen to Solve"	Portfolio Expansion "Analyze to Expand"	Opportunistic Diversion "Synthesize to Lead"
Product Users / Products	Buyers / Competitors	Workflows / Trends	Enablers / Synergies

Objects, Attributes, and Sources Table

You won't be able to investigate every object, attribute, and source all at once, all the time. Later in the book we'll discuss prioritizing your efforts, but creating an OAS inventory is Step #1.

OAS INVENTORY EXAMPLES AND THE INNOVATION SPECTRUM

As mentioned earlier, your OAS inventory will certainly be more extensive the farther across the Innovation Spectrum you are. Here are some examples:

REACTIVE INNOVATION: BUSINESS AS USUAL

Reactive Innovation is exactly what it sounds like: do what your users tell you to do. This may not sound much like innovation, but there are times in any product lifecycle where the most important goal is keeping existing users happy with what they already have. Your OAS inventory will be current product users and the product itself. If your goal is to keep existing customers happy while spending minimal resources to do so, your focus can be this tight. But what if you're looking to do more?

RESPONSIVE INNOVATION: A STEP BEYOND REACTIVE

The goal of Responsive Innovation is to anticipate the needs of product users, their leaders, and others involved in the buying process by being on the lookout for new edicts from regulatory, association, or other governing bodies that could change the requirements of your market. A Responsive OAS inventory will also include competitors and those who influence your markets, such as analysts or crowdsourced rating sites. In terms of attributes specifically, you're looking for changes being made now, as they relate to what you offer today.

So far, building an OAS inventory is fairly straightforward and probably includes the types of market analyses you're doing today. To lead—or create—an industry, your focus will expand to include objects, attributes, and sources you may not have listened to before. This transition begins with Inventive Innovation.

OBJECTS OF OPPORTUNITY RESEARCH ALIGN WITH PHASE OF INNOVATION DESIRED

REACTIVE

Reactive Response
"Listen – React"

Product Users

Products

RESPONSIVE

Additive Upgrade
"Listen to Solve"

Buyers

Competitors

Influencers

Governors

INVENTIVE

Portfolio Expansion
"Analyze to Expand"

Workflows

Trends

Markets

Gatekeepers

DISRUPTIVE

Opportunistic Diversion
"Synthesize to Lead"

Enablers

Synergies

Ecosystem

Meteors

Objects of Opportunity Research Align with Phase of Innovation Desired

INVENTIVE INNOVATION: ADDING UNDERSTANDING OF WORKFLOWS AND TRENDS

Inventive Innovation is about expansion: building new solutions for markets you already serve or offering current products to new markets to use in new ways. Looking for new markets for existing products or building new products for a market you already serve are typical actions for most growth-centric organizations, but too often we end up stuck in Reactive Innovation because we talk to existing users of our existing products, rather than prospective users in the new market or buyers in the current.

To find market-driven opportunities to grow, you'll expand your OAS inventory to find and prioritize new markets and needs. This means investigating adjacent personas and markets for relative strength and interest, understanding an entire workflow to find opportunities for expanding your footprint and the gatekeepers who may need to be overcome or included in your marketing strategy or sales process.

Inventive Innovation is also where you add the concept of *time* to your investigation. Trend information can help you further prioritize the expanded opportunities you find. Are some market problems more important than they were before? What functional role is no longer part of the buying process? Which is the strongest market that we could approach next?

This is also where you explore beyond "product" to innovate. Your big opportunity may not be in creating a new product or even revamping an existing one. Rather, it may be in a new pricing model or bundling multiple offerings into a solution. It could be selling your product online or through a partner—perhaps even embedding your product in another. Clearly, the market scope and topics of your market research and innovation strategy are getting increasingly sophisticated—and broad.

Now—the big kahuna: Disruptive Innovation. How can we possibly focus when trying to find a game-changing idea? Isn't this where chaos **has** to reign? Just the opposite—it's where you most need to focus your efforts in order to find inspiration you can act on.

DISRUPTIVE INNOVATION—A BALANCE OF BREADTH AND FOCUS

Organizations that aspire to Disruptive Innovation synthesize information from a wide variety of sources, but they keep their OAS inventory in check because they are aligned at a high level on what they do and who they do it for. Companies that build time-and-billing software for law firms may expand to additional workflow enablement software, consulting, and other services for law firms, or expand to offer time-and-billing to other industries, but they're not looking to create packaging solutions for the cheese-making industry.

They do, however, look at enablers that could come from any industry. Enablers include new technology for building products, pricing models that resonate with the markets they want to sell to, or ways to deliver a product that they aren't currently leveraging.

Disruptive innovators also look at the OAS inventory holistically to identify synergies. An eye to synergies gives disruptive innovators inspiration from how their objects and attributes relate to one another in a way that might provide an opportunity. A great example of this is how subscription pricing models have been picked up by a wide variety of organizations from video/streaming services to B2B software to legal services.

So the company offering time-and-billing software to law firms will have a very different OAS inventory depending on whether its goal is to be reactive to its market or aspire to Inventive Innovation.

Let's take a look at an example: the first section of this graphic illustrates that, to be reactive, the company only needs to listen to the lawyers, legal secretaries, and receivables clerks that currently use their existing product.

EXAMPLE: LegalcoEG SOFTWARE COMPANY

OBJECTS TO RESEARCH IF THE INNOVATION PHASE IS REACTIVE

Lawyers, Legal Secretaries, Receivables Clerks

"OurProd" Time & Billing Software

OBJECTS TO RESEARCH IF THE INNOVATION PHASE IS INVENTIVE

Lawyers, Legal Secretaries, Receivables Clerks

Time & Billing Today Podcast

"OurProd" Time and Billing Software

American Bar Association, State Bars

Managing Partner, CIO, CTO

Workflow to Track and Submit Billable Hours

"TheirProd" Time & Billing Software

Legal Industry Trends

Trends in New Billing Categories & Billing Methods

Head of Accounting

The lower section lists the much wider variety of objects this company would investigate if their goal was to be inventive and includes trends and technologies that could be applied to their offerings and the time-and-billing workflow itself.

Of course, these are both just examples of OAS inventories; you may have multiple lists from the perspective of a product, a persona, or a market as the object. The number of objects and attributes you should investigate, and the sources you should include, will depend on several factors including the phase of the Innovation Spectrum you're trying to achieve. As a general rule, the more disruptive the innovation you seek, the broader your research—and extensive your OAS inventory—will be. Later on we'll discuss how to prioritize your investigations, but first let's review how to use the OAS inventory you've created.

A spreadsheet capturing this inventory will help you move forward, but how will you capture the *imagination* of your cross-functional teams? How will you create a compelling image of what needs to be done to get alignment with—and funding from—leadership?

Now we'll answer the question: How do I align and engage my organization around the effort needed to deliver market-driven innovation? To do that, we'll build the Market Strategist's Innovation Map.

VISUALIZATION: THE MARKET STRATEGIST'S INNOVATION MAP

It's not enough to *do* the work we're discussing. For it to be useful, it has to be *shared*. Visual representations are often the best way to get across complex information quickly, so I've created the **Market Strategist's Innovation Map**, a graphical representation of your OAS inventory.

The Innovation Map for the law firm software example above could look like this:

THE MARKET STRATEGIST'S INNOVATION MAP FOR LegalcoEG'S MARKET: LAW FIRMS

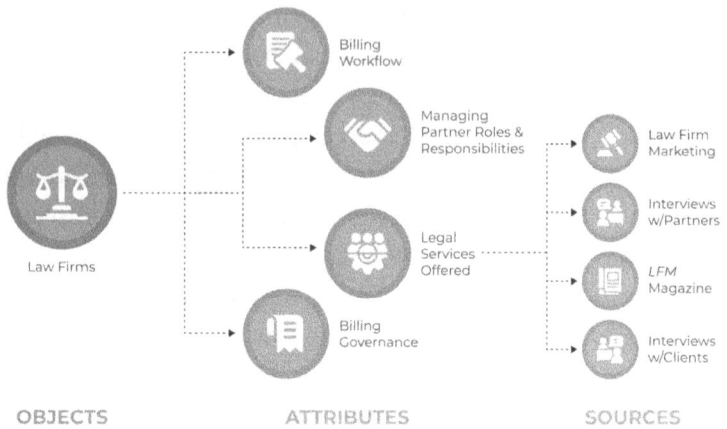

Law Firms

Billing Workflow

Managing Partner Roles & Responsibilities

Legal Services Offered

Billing Governance

Law Firm Marketing

Interviews w/Partners

LFM Magazine

Interviews w/Clients

OBJECTS — ATTRIBUTES — SOURCES

Creating an Innovation Map provides many benefits, including:

- *Enabling leadership to understand the scope of your efforts.* Even innovation on purpose can be expensive and risky; sharing an Innovation Map gives your management a tool to visualize the direction you're headed—and justification to fund the expedition.

- *Helping team members understand their role in gathering information.* For example, your IT department could be a source of information. They're most likely to know about the latest enabling technology, such as machine learning and AR, that's available to solve market problems. *Sales* expects to deliver market knowledge to the product team but not IT. With an Innovation Map, they can—literally—connect those dots.

- *Focusing the efforts of your market analysis teams.* They're the experts in data mining and analysis; giving them direction will accelerate their output and improve the results.

The Market Strategist's Innovation Map is a visual representation of the objects, attributes, and sources you and your team need to understand in order to innovate at the phase expected by leadership and to fulfill on the plan outlined in your contributing strategy. It helps the organization understand the scope and focus of your activities.

WHAT ELSE COULD YOU INCLUDE IN YOUR INNOVATION MAP?

The most important message being conveyed by the Innovation Map is where you intend to look for inspiration for market-driven innovation. However, you could add insights to the graphic as you progressed, such as:

- *Timeline for research.* Adding a note for which month or quarter you expect to investigate each source helps you manage expectations around when you'll have recommendations for new directions.

- *Depth of current knowledge.* It's impossible to always know everything about every object. Providing a graphic indication of the depth of your current knowledge and confidence in your sources increases your credibility and ensures leadership knows the level of risk you're taking when decisions are made. I mentioned creating a Confidence Language in Chapter 4; as I said then, we'll expand on this topic later on.

- *Relationship of objects.* Across product portfolios or markets, objects such as personas may play different roles or have different needs. It could get a little messy, but bringing all maps together in one visual can drive home the complex nature of market-driven innovation.

Searching for innovation opportunities without a strategy to guide you is frustrating, inefficient, and expensive. To innovate on purpose, you must look for *inspiration* on purpose. That means building an Innovation Map of objects, attributes, and sources that focus you on finding insights to take you in a direction you want to go.

ACTION PLAN FOR CHAPTER 6

- Make sure you know who in your team is responsible for creating and maintaining knowledge structures.

- If it's you, refer back to your Innovation Spectrum and contributing strategy to draft a Market Strategist's Innovation Map for one of your products. Review with your leader for edits and approval.

- If another team (e.g., data scientists, market research team) owns this activity, meet with them to share your Market Strategist's Innovation Map and brainstorm how to ensure you can access the information you need on the objects your contributing strategy requires you to understand.

Chapter 7
AGGREGATE WHAT YOU ALREADY HAVE

Organize Current Knowledge

- Aggregate What You Have
- Create Knowledge Statements & ID Gaps
- Prioritize Filling of Gaps

The next item in the Organize Current Knowledge step is to map the market knowledge you already have into the data structure you defined in the last chapter, in order to enable access to these different sources and formats in a way that allows you to understand your markets. In the absence of a taxonomy already in place in your company, I recommended structuring with objects, attributes, and sources.

In my experience, lack of market knowledge aggregation in order to find patterns and trends is where most organizations are leaking a lot of the value they could be getting from their current market research. They gather huge amounts of market data through surveys, focus groups, and data mining, but they only examine and use it at the moment it's gathered—they gather, review, act (or not), and then throw it out. Worse, some market insights are gathered and never used at all.

The aggregation workflow I propose is intended to be as easy and straightforward as possible, but it *does* require systematic attention, some allocation of time, and a plan. In this chapter I'll lay out exactly how to get all your market data in one spot by creating a single source of truth—a Market Strategist's Knowledge Center.

> **Your Knowledge Center is the Market Strategist's single source of truth about the market. Wherever market knowledge is stored, the Knowledge Center is the toolkit through which you access and analyze the data.**

A Knowledge Center requires three steps: deciding how you want to store and access market data, gathering what already exists, and plugging in ongoing data streams.

CREATE A MARKET STRATEGIST'S KNOWLEDGE CENTER

Create a Market Strategist's Knowledge Center

CHOOSE SYSTEMS AND TOOLS	AGGREGATE WHAT EXISTS TODAY	PLUG IN ONGOING SOURCES
• Innovation team requirements • Functional taxonomy • Assist in choosing software (Aha!, productboard) or DIY tools (Excel, Dropbox)	• Scavenger hunt • Ensure someone owns aggregation and does it carefully • Add most recent first	• Internal • External • Data mining

CHOOSING SYSTEMS AND TOOLS: REQUIREMENTS FROM THE MARKET STRATEGIST

Remember my caveat from the last chapter about my not being a data scientist? I'll add another: I'm not a software designer or BI system specialist either. And, just as in Chapter 6, I want to emphasize that *building* the single source of market truth is probably not your role. If you have a market research, data science, or business intelligence group, they're almost certainly doing this work or already have it done. Don't build a separate system!

However, just as I mentioned in Chapter 6, you *are* the primary user of this information and must provide whoever owns the technical side of data aggregation with *business* requirements so you can achieve the market understanding necessary to innovate on purpose. That will be the primary focus of this chapter.

What if you and your research team are just starting out and are looking around for tools to help? The number of sophisticated out-of-the-box software tools to aggregate and analyze market data have exploded in recent years—and the options keep growing. If you have the funds to buy or build a sophisticated system like Aha! or productboard, good for you!

No budget? Don't wait for the long process of approval, evaluation, and implementation—that could take years. Instead, brainstorm with your tech team and analysts to utilize the tools you already have: Excel, Dropbox, Teams—whatever you're already paying for. It would be nice to have software that did a lot of the aggregation work for you; but even manual, patched-together systems are vastly better than nothing. Too many Market Strategists wait and hope to get budget and approval for the "dream" system. Don't wait—start with what you have.

Whatever tools you use, your goals are to be able to access and analyze your market data in ways that allow you to find inspiration to innovate according to the goals you've aligned on with leadership. So rely on your internal experts if you have them, but make sure you provide them with the business requirements for any tools implemented. Following are a few suggestions for business requirements:

- *Integration of information from other internal systems such as Salesforce, JIRA, and others.* Ideally, your tool automates the call to whatever back-office or workflow systems you need to tap into. Your website traffic tracker, lead-generation software, sales and customer-service workflow software—these are obvious tie-ins. But don't forget online help/chatbots, email, texts, and phone calls of all teams interacting with the market.

- *Access information from* **external sources.** Alerts, public-record data, or other sources outside your organization hold valuable insight. Where possible, enable automated gathering.

- *Manual entry of information received offline.* Most tools are good at aggregating information from other systems but give less thought to the valuable *offline* sources. An engineer sees something cool at a trade show. You have handwritten notes from a focus group. A salesperson hears about a new competitor. You want to keep all of this insight, so make sure the system you're using allows you to manually enter them.

- *Flexible/customizable data structure.* Whether you use objects, attributes, and sources as your data structure or one that your organization already uses, be sure your tool can handle your specific classification needs.

- *Associate data with one or many objects and/or attributes.* A survey may provide insight both into the managing partner persona as well as the overall law firm market. It's important to be able to leverage all market data wherever it makes sense to.

Once the system you're creating can bring all the market knowledge together in one pile, there are many insights you'll want to gather from it. Be sure you're able to do the following types of analysis:

- *Multi-perspective.* You may want to see all data on a single object, or only the data about a specific attribute of that object, from a single source. You may want to see data since you began collecting it, or only from a particular period of time. It's possible you may even want to search for the presence of a keyword. Look for a high level of flexibility to mine data if you're buying a system.

- *Qualitative and quantitative.* You will be using your toolkit in many ways, including the need to pore through narrative data for patterns and identifying the need for follow up. AI and machine learning are making this easier than ever, but dive deeply into the capabilities and limitations of your system—you may want to do more or different reviews than your bot allows. Make sure you can.

- **Strength of individual insights.** Not all market knowledge is created equal. Some sources are better than others, and some data reflects more of the market than others. Can your system score a data source? Can it score the source differently when it's linked to different objects and attributes?

- **Strength of aggregated insights.** This is the ability to create a cumulative "strength" score on any given object, object/attribute, or source of data; more on this later.

- **Trend analysis.** Where you capture the same date over time, such as open rate on marketing emails or use of particular product features, ensure that your tool has a way for you to analyze changes over time. Past behavior isn't a perfect bellwether of future behavior, but trends can give you a sense for where your markets are headed.

The analysis you do now helps you innovate on purpose immediately, using the hard work you've already done on market research. But you're also going to find gaps in your knowledge, so I have three more recommendations to ask of any system you implement:

- **Identify knowledge gaps.** Some way to visualize a comparison of the market knowledge you have with the objects and attributes you need to have knowledge of.

- **Prioritize the gaps.** A major gap in most innovation workflows is that there's no innovation *plan*. Just as product managers prioritize feature builds and then chunk the work into sprints or other iterations to execute, so you must also prioritize your market analysis and innovation.

- **_Generate a prioritized market backlog._** Whatever tools you use should enable you to create (and, if necessary, change) a prioritized backlog of blind spots in your market knowledge that you'll fill over time.

Every organization is different, and this isn't an exhaustive list. If you have them, work with your engineers and data scientists to create a system that's right for you. And remember: if you can't afford a specialized software solution, use what you have.

Completing this step is a big milestone for most Market Strategists. Now you can aggregate the market insights you already have, which means you can _use_ the market insights you already have. Take a systematic approach to gathering up what exists today. I call this step the scavenger hunt.

AGGREGATE WHAT EXISTS TODAY

One of the most difficult parts of building a Knowledge Center is populating it, both initially and consistently. The good news is this is almost never for lack of documented insights. How many surveys have you done in the past year? How many data sources do your scientists have at their disposal? How many sales calls, customer service, or chatbot interactions are stored? Chances are, you already have a lot of actionable information from a variety of sources that show clear trends to either investigate farther or act on now. Don't waste these! Gather them up to find opportunities that may already exist. It's time for the scavenger hunt.

GO ON A SCAVENGER HUNT

Scavenger hunts have been popular at kids' birthday parties and corporate events for decades. Pick some teams, give each a list of random items to find and—first one back wins! A scavenger hunt to innovate on purpose, on the other hand, is a gift that keeps on giving after the hunt is over. The simple act of gathering up the market knowledge that exists today and considering it as a whole against your responsibility to achieve specific goals is one of the best uses of a couple of hours that I can think of.

When you perform a scavenger hunt for existing market knowledge, you leverage research you already paid for. I remember the pull-and-tug of teams trying to get a budget for new initiatives, and me saying, "Give me something to react to, then I can decide how to fund this." Using what you already have on hand to find opportunities to innovate is the ultimate in upcycling.

The scavenger hunt also gives you a chance to validate your OAS inventory and other data structures without waiting around for inbound data. I once discovered we had a company subscription to a valuable analyst report nobody in my department was leveraging, because we didn't know it existed.

A scavenger hunt allows you to find opportunities and get results right away. I've always hated plans and workflows and processes that didn't deliver something at the get-go. The scavenger hunt breaks that mold.

And the reason it breaks the mold is because you **will** find something new and promising. It might not be game-changing. It might not even bring in new revenue. But then again—it might. I once had a scavenger hunt uncover a batch of "customers" whose free trial of a product they used like crazy had never been converted to subscriptions. We picked up six figures in new—recurring—revenue with that little discovery.

Following are some tips to keep in mind when going on a scavenger hunt:

- *Make a plan.* It's inefficient to just start wandering around asking your colleagues, "Got data?" You need a plan.

  From your OAS inventory, you have a checklist of departments you need to hit (sources). Be sure to include not just sales and customer service but legal, accounts receivable, implementation teams, sales specialists, market research, and data scientists. If you already have your data wrangler role in place, they can coordinate the project. You may be surprised at the research or data you find hiding in plain sight. Ask other departments to review your source inventory and suggest anyone you missed.

- *Have leadership announce.* A well-timed communication from leadership would be nice, too. If you can, get your executive to send a general announcement to the organization explaining your efforts and asking for everyone's cooperation. Market Strategists who have this kind of air-cover get a lot more cooperation a lot faster than those who don't.

- *Choose a timeline for assets to retrieve.* It's not practical or useful to add every survey or focus group you've done since the beginning of time, so it will be necessary to set a cutoff date. Whether it's three months or three years, choose what makes sense for your organization. While I usually recommend going back 18–24 months, you should take into account the volatility of your industry and markets (including indirect influences such as supply chain issues or significant competitive entrants.

- *Prepare for the discussion.* People want to know why they're being asked for something. Why do you need my research? What are you going to do with it? Be ready with your reasons, and an example or two, of why having all market research consolidated would be valuable. "Please" and "thank you" will also go a long way. So does making your request through a medium that the person is comfortable with. If they prefer Slack, communicate through that channel. If they'd rather speak in person, walk over and ask. Afterword, follow up with a thank-you and some detail on what you gathered and why it was valuable. Finally—go public with your appreciation. "The study Onika in sales gave us was the best evidence of…" reinforces that cooperation is not only valuable but appreciated.

- *Accept "no" for an answer.* We live in the real world, and I know from personal experience that not everyone you ask will be willing to hand over their market research, even when you give them a good reason to. Building trust and defusing skepticism is a long-term activity, and you won't achieve it at this moment by making a stink or turning the request into a power struggle. Don't argue (even inside your head) with those who won't give you access to their research. Whatever their reasons, you have a lot to do and can't afford to get sidetracked. Over time, those data gaps will become obvious, then awkward, and others will wonder why those folks aren't sharing with you. You can wait. Remember, right now your Knowledge Center is empty. Whatever you add will be valuable.

- *Make it easy.* You may be aggregating this knowledge for the good of the organization, but it's still *your* project. You're asking others to help do *your* work (yes, that's how they'll see it). You'll be more successful if you make it as easy as possible for colleagues to give you what you want. If all they have are hard copies of research, scan and return them promptly. If it's recordings, transcribe them yourself. Say please and thank you!

- *Remember external knowledge sources.* Depending on the phase of innovation your organization is looking for, you may also need to plug in external data sources such as regulations, partner surveys, or reseller data. Don't forget to aggregate what's available there to add to your baseline.

CONSOLIDATE WHAT YOU FIND

Great! Now you've gathered the historic data you want to retain. And, in its current form, it's useless. To identify trends and patterns you need to get this information into your Knowledge Center/ aggregation system. It's a huge job—where do you start?

- *Begin at the end.* While I recommend gathering research as far back as makes sense for your organization, begin your input with the most recent and work backward. There are two main reasons for this. One, recent market research is generally more immediately actionable than older research, which you're largely inputting to provide trend analysis. Two, somebody paid for the recent research recently and may have a higher interest in seeing you act on it than something done 18 months ago.

- *Input carefully.* This is where the data wrangler is particularly valuable. The data wrangler is part of the process and, with the IT team, market researchers, and data scientists, responsible for accuracy and utility of the data. Specifically, the data wrangler is primarily responsible for accurately and exhaustively tagging and linking the manually entered knowledge into your system and checking the electronically added data for quality also. It's more important to be thorough with tagging and organizing than it is to get the assets aggregated. Why? Because even when everything's accessible through one interface—you can't use what you can't find.

- *Make aggregation a priority.* Doing more and more research that you don't aggregate and can't analyze doesn't make sense. So particularly if you've done a lot of market research that you're not using, shift focus to aggregation for a while. As you move forward, allot a certain amount of time each week for data input and quality control.

- *Tweak your data structure.* As you bring these disparate sources of research together, you'll probably find some objects, attributes, or sources you didn't account for in your initial data structure. Add them as you come across them but remember: no data structure or system is perfect. Improve as you can, but start where you are, and use what's available today.

I'll never forget the faces of one of my teams as they came back from their scavenger hunt with binders and thumb drives and recordings. It was as if they'd found treasure hidden in the nooks and crannies of the organization. They had.

My team and I were sifting through our newly aggregated data. During the process, a couple of team members from marketing and customer service approached me with a *terrible* data point—95% of prospects who called customer service with questions about buying online were walking away *without* buying!

We knew our online, transactional buying workflow was less than stellar—our main business was annual licenses. The transactional purchase workflow was really complex, with lots of if/then options to navigate through when making even a simple purchase. Further muddying the waters, our shopping cart was homegrown, and didn't operate like a standard shopping-cart workflow. Topping it all, our policy was to refer confused shoppers to the online shopping user manual—an outdated, hard-to-use PDF that was not very helpful.

The two proposed that customer service assist the callers by filling online shopping carts for them and then emailing the cart link for payment and approval. A very patched-together, low tech test, but we tried it.

The results were staggering. More than 70% of the walkaway traffic stayed to buy when customer service filled the cart and emailed it to them for final approval and payment. Many came back to shop *again* and considered buying licenses. The project that cost exactly zero in incremental budget increased profits by almost a million dollars—ongoing!

Not all knowledge will be so lucrative, but there will be gems hiding in the work you've already done. Finding them is the first big proof-of-concept you'll have to increase everything from awareness to budgets.

As you executed your scavenger hunt, you probably identified sources of knowledge you should listen to on an ongoing basis:

information streams that deliver the continuous market understanding you need to fill in the gaps of your Listening Map and achieve your contributing strategy. We'll create that workflow that now.

PLUG IN CONTINUOUS AND FUTURE INFORMATION STREAMS

The scavenger hunt gives you a baseline of information: using what you already know to innovate right now makes sense! But your markets are always changing, and new data is always flowing in. To innovate on purpose, you'll have to keep up with those changes.

The good news is, if you've invested in a purpose-built tool, this work will be less labor intensive than the last step, for the electronic sources anyway. If you bought Confluence or another system, plugging in automated sources of data from workflow-enablement apps such as Salesforce will likely be your first step. But while automation will take care of electronic data sources, you need to ensure that ongoing "offline" sources of information are also getting into your Knowledge Center. And we've already discussed the reality that finding innovation opportunities beyond the Reactive phase often requires gathering information beyond customer and product usage data, information which is often not available for auto-integration.

The questions I most often get at this point are, "What sources do I need to gather?" "How am I supposed to get this information?" "How can I make anyone help me on an ongoing basis?" Let's explore each of these:

To determine what sources to gather on an ongoing basis, you should return to your Innovation Map. You've already listed the sources you need to get information from on the map (and in your OAS inventory). The next step is really a to-do list to get them integrated into your Knowledge Center.

Even if sources are automatically streaming into your software, it's good to have a contact who knows not only how that's happening (usually an IT team member) but also the background on the data: where it's actually coming from, what its strengths and weaknesses are, and how often it's updated. Once you know that, you can plan to integrate learnings into the Knowledge Center whether it's unstructured insights from an industry blog or quantifiable responses from a survey.

Below is an example of what this could look like for LegalcoEG:

LegalcoEG: INFORMATION AGGREGATION PLAN

SOURCE	INTEGRATION TACTICS	CONTACT	FREQUENCY
Advisory groups, interviews, peer-to-peer interviews W/L analysis.	Electronic transfer when possible; mostly scan, transcribe, manual load.	Kavita in market research.	She'll send last week of each month.
Monitor press for competitor, regulation, market and product mentions.	Most come from our Prowly subscription; some manual.	Rowan in PR will enter and add images, weekly as received. Also the contact for all electronic feeds.	Weekly and automated feed.
Customer service inbound.	Calls: recordings kept. Bot: electronic integration from bot. Escalations: electronic through Salesforce feeds.	Dan, Director of Customer Service, will send. We need to review and enter reasons into our notes.	Sample of 25 loaded to Teams channel weekly; automated feeds.

THE MARKET STRATEGIST'S INNOVATION MAP FOR LegalcoEG'S MARKET: LAW FIRMS

LegalcoEG: Information Aggregation Plan

IT TAKES A VILLAGE—BUT IT'S YOUR RESPONSIBILITY

Remember that any information someone sends you is extra effort above and beyond their day-to-day work. Salespeople are there primarily to sell. Customer Service, to help customers. And for all the "work together" messaging from leadership, these individuals are compensated for meeting *their* goals, not *yours*. External sources may not be even indirectly impacted by your success. So make it easy for them. Set up a dedicated voice mail, Slack or Teams channel, email address. Allow your colleagues to text information. Any way your company communicates, enable that channel for gathering data. Don't force a form or a lot of rules into the process—just get the data.

Plug in your ongoing sources of information one by one. This will go quickly if you're a Reactive Innovation team with an automated system, and slowly if you're trying to be disruptive with a homegrown set of tools. Make a plan and remember—whatever you get into the Knowledge Center is more aggregated insight than you had before.

The steps we've covered so far will likely give you plenty of inspiration to innovate on purpose, whatever your innovation goals are. Don't wait to leverage your new insights until you have a perfect Knowledge Center—start now! You already have more actionable insights than you did before. As a Market Strategist you'll advocate for continuous improvement in your connection between market knowledge and innovation, but you won't hold up the organization because you don't have "perfect" knowledge.

That said—you still have knowledge gaps, but you always will, right? When is enough market data, well, enough? How do you know when you have enough information to *act*? Let's discuss how to get to that step now.

ACTION PLAN FOR CHAPTER 7

- Touch base with your data science, IT, and market research teams to see if you already have an aggregation system in place. If so, make sure it meets your requirements to understand the objects, attributes, and sources on your Innovation Map.

- If there is no system in place, brainstorm with this group to get their advice on what in-house tools could be used to enable aggregation and analysis, even if they're not purpose-built for that work.

- Go on a scavenger hunt to gather the market insights that exist today. Before you begin, make a list of departments and people to see; ask them where else to look.

- Set aside some time every week to enter data that doesn't stream into your system and begin with the most recent research first.

- Plug in continuous data streams so you can keep up with market changes.

Chapter 8

CREATE KNOWLEDGE STATEMENTS & IDENTIFY GAPS

Organize Current Knowledge

- Aggregate What You Have
- Create Knowledge Statements & ID Gaps
- Prioritize Filling of Gaps

The reason you gathered this data into a pile is to discover insights about your markets and personas so you can innovate on purpose and succeed in the market. But action requires the confidence of your organization. How do you make them comfortable that you have *enough* knowledge to act? How do you telegraph relative risks due to the ever-present reality of incomplete knowledge and how should you prioritize which of those knowledge gaps to fill with your limited time and resources?

Let's answer those questions now.

In this chapter, we'll begin translating data into knowledge, identify knowledge gaps, communicate confidence levels, and lay a foundation for prioritizing the additional data-gathering you need most to innovate on purpose.

TRANSFORM AGGREGATED DATA INTO KNOWLEDGE STATEMENTS

Creating **Knowledge Statements** is the task of reviewing your aggregated data and articulating what you already know about the objects and attributes on your Innovation Map.

Knowledge Statements aren't product requirements or messaging for the market but rather a short sentence or phrase describing something your aggregated data says about the object. You create them so you and your team can understand the objects of your research; Knowledge Statements transform one or several pieces of data from multiple sources into a bite-size statement.

> **Knowledge Statements are short sentences or phrases that aggregate one or many pieces of data from your Knowledge Center into actionable statements about an object. The Market Strategist creates these and communicates them to the rest of the team so everyone can do their jobs well.**

Let's once again use our LegalcoEG example. If your object is the market of law firms like LegalcoEG, you're going to want to know the key challenges law firms are facing now and those they feel will be their biggest challenges in the future. "Biggest challenges" is an attribute of the object "law firm market."

Because you've done your scavenger hunt and aggregated all the information you currently have on the law firm market, you can review that information and, likely, document a slew of Knowledge Statements for any given object on your Innovation Map.

Maybe you did a survey about key challenges a couple of months ago. Maybe you also bought an analyst report that summarized anticipated challenges law firms think they'll face in three years. Maybe a couple of sales reps have come back from a quarterly check-in and shared some updates.

In the example, three Knowledge Statements came out of reviewing the data in the Center: *"The top challenge for law firms today is billing," "Law firms expect hiring to be the top challenge 3 years from now,"* and *"Law firms don't see a good hiring tool in the market."*

If you've got aspirations for Inventive or Disruptive Innovation, you're probably wondering where to begin with all the objects and attributes on your Map. Especially with these complex Maps, creating

Knowledge Statements is going to be time consuming! To keep your sanity, I recommend tackling it one object and attribute at a time. Prioritize the work based on the objects that most significantly impact your ability to achieve your contributing strategy. I used to prioritize this work by asking myself, "What object and attribute do I know the least about and worries me the most?"

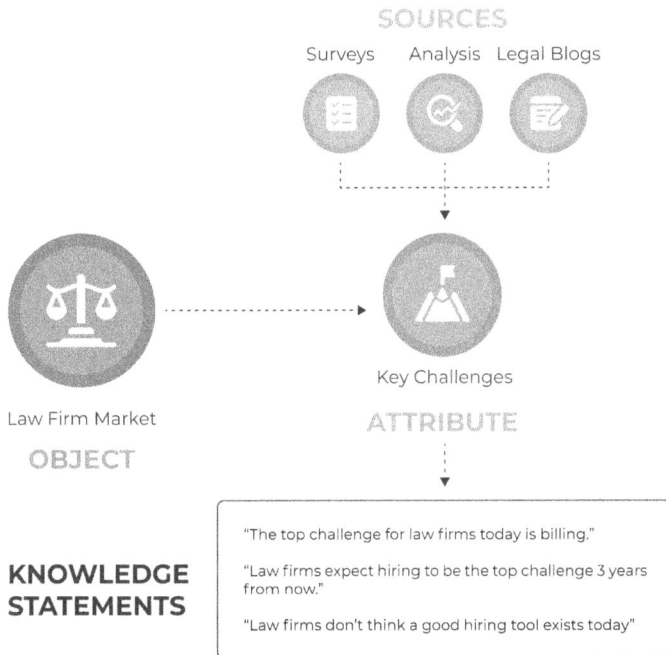

CREATING KNOWLEDGE STATEMENTS

SOURCES

Surveys   Analysis   Legal Blogs

Key Challenges

Law Firm Market

OBJECT

ATTRIBUTE

KNOWLEDGE STATEMENTS

"The top challenge for law firms today is billing."

"Law firms expect hiring to be the top challenge 3 years from now."

"Law firms don't think a good hiring tool exists today"

Creating Knowledge Statements

If you're trying to open the Canada market, as LegalcoEG is, perhaps you start with "Managing partners of Canadian law firms" as the object and "Concerns about time and billing" as the attribute. Don't make yourself nuts trying to do everything at once, but don't keep gathering information you don't use, either. Set aside whatever time you can to do this work, even if it's 15 minutes each week. It will take longer to get completely on top of your Innovation Map, but you'll have more actionable information after the first 15 minutes than you do today.

The good news is, as you discover these patterns, you'll likely find that you have an enormous amount of knowledge, substantiated by a wide variety of sources, that confirms current assumptions about your personas or markets. Even better, you will certainly find even more opportunities for innovation that you can pursue right now, without gathering any more data.

Some Knowledge Statements will be less well documented than others but nonetheless be actionable if you're willing to take some risk. Some "knowledge" your organization has set down as true for years will have no basis in data at all.

Now that you're building statements your organization might act on, it's important to create a method to signal how confident you are that the statement *is* actionable, based on the amount and quality of data you've gathered.

CREATE A CONFIDENCE LANGUAGE

The terms "fact" and "true" are so loaded these days and so often bent to prove a predefined result, that I would stop using the terms altogether. Instead, create a **Confidence Language** based on tiers of beliefs, data, and knowledge, to determine which Knowledge Statements are actionable or not.

> **A Confidence Language establishes a series of levels that telegraph how confident you are that your Knowledge Statements are true. The Market Strategist establishes levels based on such criteria as source of data, recency of reporting, and a sense for the risk tolerance of their own organization.**

Telegraphing how confident you are in a Knowledge Statement is an excellent way to understand risk, align teams, and get the budget you need to innovate. Back in Chapter 4 "Aligning with Leadership,"

I referenced building a Confidence Language so you and leadership could get aligned on how much market research you should do—how much knowledge you should have—before you spent resources to respond with market action. You can create whatever system you choose, but the five tiers I suggested in Chapter 4 are the ones I'll define and expand on now, with the first three being what I would consider mandatory tiers:

- *Believed*. An often historical viewpoint believed by one or many in the organization to be actionable knowledge. It's one opinion, but there may be others. E.g.: "Customer service believes their service is better than competitor A because they work hard to *offer* excellent service, but product management disagrees, and we have no data that confirms either position." Believed knowledge is often shared after the phrase, "*Everyone* knows...." Believed knowledge *may be* true, but there's been no market research or documentation of individual experiences to *ensure* that it is. The weakness in Believed knowledge is that, without documentation, it's difficult to get funding or rally a team around it. Believed knowledge represents the highest level of risk.

- *Perceived*. An assertion based on limited but sound, documented experience and at least some market data.

   A good example of this is the conversation Market Strategists often have with members of the sales team who bring in a one-off contract. These are sometimes positioned as being representative of the market as a whole, when in fact the only point of data from the market is the discussion with this single prospect. "You know if they want it, everyone will want it!" is a discussion that often stems from overreaching with Perceived knowledge. Perceived knowledge could also come from one or two customer services calls, a survey response, or other quantifiable, but isolated, piece of information.

   Acting on Perceived knowledge is less risky than acting on Believed knowledge, but it's better to conduct additional experiments or market research to further validate.

- **Recognized** knowledge has reached the threshold set by your organization that the knowledge is **actionable**—you can and should use it to shape market responses. Recognized knowledge has been quantified, or heard from specific sources that are considered reliable, or from enough sources to establish a representative sample of the market.

   For example: "*AnalystBiz Report* recently ranked us #1 in customer service. Our own survey shows the law firm market ranks our customer support team #1 in industry knowledge. Win/Loss analysis shows the #2 reason customers renew is "Excellent service." This is Recognized knowledge.

You could also create two additional designations to add more nuance to your system:

CREATE A CONFIDENCE LANGUAGE

Create a Confidence Language

Telegraph level of certainty to the team.

Establish common understanding.

- Do we have enough knowledge to act?

- If not, what else do we need to feel comfortable?

- Create a list of hypotheses to validate.

Establish a language for market change.

UNKNOWN (GRAY)
"We should know this, but don't."

BELIEVED (RED)
"We believe this."

PERCEIVED (YELLOW)
"We have some data on this."

RECOGNIZED (GREEN)
"This meets our data threshold."

TRANSITIONING (BLUE)
"This is changing."

- **Unknown** can be defined as the blind spots in your market knowledge. For example, if you as the Market Strategist are responsible for trial-preparation software solutions for the

North American market and you're unfamiliar with the restrictions on expert witnesses at trials in New York state courts, this would be a designated unknown. Unlike the "We don't know what we don't know," these are knowledge gaps you know about and should at some point fill.

- *Transitioning.* This is knowledge that was formerly Recognized, but there are indicators of change. E.g.: "Our customer service NPS scores have ticked down in the past year; it's possible we're losing our edge." It could also mean that you've not checked in to reverify Recognized knowledge in a period of time that makes it suspect or that there has been an event, like the COVID pandemic or a natural disaster, that makes that knowledge suspect even if that time has not elapsed.

Leveraging a common language for defining what you know about your markets helps telegraph to leadership, teams, and new hires your relative certainty that something said about a market is true. Action can be taken based on this common understanding and level of risk.

DEFINING CONFIDENCE LEVELS

The next step to an effective Confidence Language is to decide what qualifies a Knowledge Statement to meet each criterion. In other words, what does it mean for an insight to be "Perceived"? What data sources qualify Knowledge Statements for that rank? How many sources are needed? There are several perspectives from which you could create these thresholds, including:

- *Strength of source.* Individual opinion, collective experience, market surveys, analyst reports; which of these is most reliable? While the latter two are usually more reliable than the former, base a strength-of-source score on what makes sense in your world.

- *Number of sources.* Another commonly used scoring method is to add the number of sources together, or the "strength of source" scores, with the logic being the more sources that say something, the more likely it is to be a strong insight.

- *Recency of data.* Is this a new data point or is it a couple years old? We'll talk later about the value of trend analysis in innovation, but generally newer data should score higher than older.

- *Proximity of data.* Too often, organizations ask product users whether they'd buy or pay for something new. Too often the users say yes, only to have buyers say no. This happens even when the buyer and user are the same person! Give extra points to primary-source data.

- *Validity of sample.* Was the cohort surveyed sufficient to deliver reliable result? Is this a count of inbound calls versus proactive market research? However you quantify sample size, it's important to acknowledge that there's truth in numbers.

While I recommend keeping your Confidence Language simple, with no more than a few criteria used to assess confidence, it may help set and use the levels if you quantify them with a scoring system.

CONSIDER A CONFIDENCE SCORE

If you bought an app to help you innovate, chances are it came with a predefined scoring system or rules you could customize. Whether you build your own scoring system or use a tool, it's still important to implement scoring based on your own organization's goals around innovation, realistic abilities to gather data, and overall tolerance for risk. In other words: an out-of-the-box score will give you out-of-the-box results.

> **A Confidence Score adds an element of quantification to the Confidence Language, by assigning a numerical value to criteria such as source of data, recency of reporting, and a sense for the risk tolerance of their own organization.**

Here are two examples of using the same data to create a Knowledge Statement that is scored differently because the two companies have different thresholds for risk-taking:

DIFFERENT SCORES, SAME DATA REFLECTS EACH COMPANY'S UNIQUE RISK TOLERANCE

ATTRIBUTE
Key Challenges

"Law firms expect hiring to be their top challenge 3 years from now."

KNOWLEDGE SOURCES AND SCORES

	#1: 2-Year-Old Survey	#2: New Analyst Report	#3: Legal Blog Last Week	#4: Other Factors	TOTAL SCORE
Company #1: LegalcoEG (High Risk Tolerance) "Believed" = Score >10 Uses only value of source	5	4	2	NA	11 "RECOGNIZED"
Company #2: LegalEGco (Risk Averse) "Believed" = Score >15 & recency of source	1	5	4	3	13 "PERCEIVED"

Different Scores, Same Data Reflects Each Company's Unique Risk Tolerance

As you can see, the first company, "LegalcoEG," is risk tolerant; their scoring system considers the Knowledge Statement to be Recognized. Based on the same data, the more risk-averse "LegalEGco" is only willing to consider it Perceived.

IMPLEMENTING YOUR CONFIDENCE LANGUAGE

Building a scoring system for your Confidence Language is important, but how you implement it is critical too. Following are some considerations to keep in mind:

- *Keep it simple.* Complexity adds nuance but also effort to any scoring system, even those that are automated. You're also less likely to consistently have complex data available. Keep your levels to five or fewer, and your scoring elements the same. If you don't have software or a reasonable number of resources, don't bother scoring.

- *Define to align.* Don't skip the definition steps. Frustration— and extra meetings—occur when you implement language without precision around what you **mean** by it. Define your confidence levels and be precise about the thresholds for meeting each of them.

- *Separate creation from use.* Create your Confidence Language without reference to any one initiative. Every insight has **some** justification for considering it; establish a system that drives the strategic goals you want to achieve and use it—in that order.

- *Take risk into account.* The farther down the Innovation Spectrum you go, the likelier it is you'll have fewer, weaker data points that are more open to interpretation. In other words, confidence may be lower overall if you're aspiring to disrupt rather than react.

- *Involve the team.* Share with your teammates and leadership and encourage discussion and critique. Getting buy-in for the idea of a Confidence Language and your methodology is an investment in future success.

- *Use consistently.* Scoring must be uniformly implemented across all Knowledge Statements at any given time for it to be useful in assessing comparative risk. That doesn't mean you *can't* modify the system in the future. But when you *do* change it, all Knowledge Statements must be re-scored.

USING YOUR CONFIDENCE LANGUAGE TO EXPRESS KNOWLEDGE OF OBJECTS AND ATTRIBUTES

You can use your Confidence Language to quantify individual Knowledge Statements, or entire attributes and objects, so everyone in the organization is aligned on what you know about your markets—and what you don't. You'll want to keep these insights in an aggregated format: a database or spreadsheet or list, but a great way to visualize the accumulated market knowledge at a high level is to color-code your Innovation Map using the green/yellow/red construct.

This part of the process can be surprising—you may find that your organization widely believes certain things to be true about your markets for which you have no supporting data. But it can also be rewarding—you may have more actionable knowledge already in your hands than you believed. This baseline also helps you identify the most important gaps in your knowledge. As you gather more market information, you can upgrade the status of your objects, attributes, and Knowledge Statements. Even if you don't have an updated visual, the ability to say, "We could act on that, but it's 'red' data" can give you the power to quickly convey—and align the team—around any risk you might be taking.

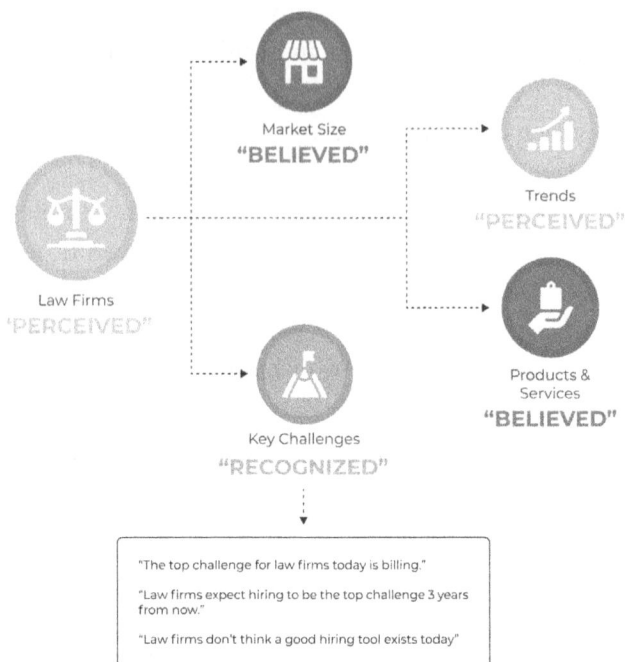

Believed, Perceived, and Recognized Innovation Map

Market Size
"BELIEVED"

Trends
"PERCEIVED"

Law Firms
'PERCEIVED"

Products &
Services
"BELIEVED"

Key Challenges
"RECOGNIZED"

"The top challenge for law firms today is billing."

"Law firms expect hiring to be the top challenge 3 years from now."

"Law firms don't think a good hiring tool exists today"

DOCUMENTING KNOWLEDGE GAPS

Chances are, you've discovered some terrific, actionable insights by now. Hopefully enough to align your organization around market needs and act with more confidence. Don't wait to use the knowledge you've identified! But you've almost certainly also discovered gaps that must be filled to unlock market-driven innovation opportunities.

Where is your Innovation Map showing a lot of gray, red, or yellow? Companies aspiring to Inventive or Disruptive Innovation usually discover that they know a lot about user personas and competitors but nothing at all about governors and enablers. They haven't been tracking trends and synergies. In short, they find that they've *aspired* to be inventive but *worked* to be reactive. To innovate, you'll need to turn these gray areas green!

Even organizations with more modest innovation goals might find they've been making decisions based on "knowledge" that everyone assumes is true but nobody's actually verified. Too often what gets stable, successful companies in trouble is that they still believe something that they learned about their markets 10 years ago that hasn't been true for the last five.

Either way, unknowns are dangerous, and the next step will be to see what we can do about them.

ADD UNKNOWNS TO YOUR KNOWLEDGE STATEMENT INVENTORY

The first step is to document what's unknown with the opposite number of a Knowledge Statement: a **Knowledge _Gap_ Statement**. Let's take a look at an example:

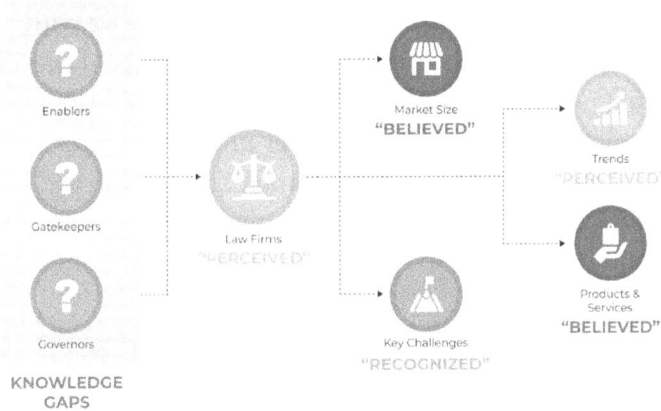

As you can see, this organization has at least some actionable insights about its market, law firms in Canada. But the gaps in their knowledge could hinder their ability to innovate.

> A Knowledge Gap Statement is the opposite of a Knowledge Statement. It documents something you don't know about an object, but should. Creating a list of Knowledge Gap Statements is the first step toward prioritizing next steps for market research. Knowledge Gaps also telegraph to the team that there is more to learn about the market.

When they put together their Innovation Map, it's possible they thought, "Hmmm … I know there are regulations governing how Canadian law firms document billable time, but I have no idea who's **responsible** for that governance." That gap is an unknown at the Map level (objects), and is represented as the lower of the three gray circles on the left, as "governors."

If they haven't kept up on the accounting rules Canadian law firms follow, their software may not enable all the necessary workflow steps their customers are required to perform.

It also looks as if they haven't learned enough about the OEMs they sell their product through (gatekeepers); this ignorance may be creating a dangerous perception of indifference with those OEMs— the very partners who deliver the revenue! Being unfamiliar with cutting-edge technologies and emerging pricing strategies of the world in general (enablers) could hinder this example's ability to graduate from a player to an innovative leader in the market.

Finally, one of the market attributes, "market size," is believed to be about half the size of the South American market, but nobody's validated that number. I'd hate to commit to revenue without checking that!

Once you've added unknowns to your object and attribute inventory, you can create baseline Knowledge Gap Statements to guide your next steps. Here are a few examples:

"We don't know what regulatory agency/agencies set discovery documentation rules for Vivienne, managing partner of a law firm in Canada."

"We don't know what the discovery documentation rules are for Vivienne, managing partner of a law firm in Canada."

"We don't know whether Vivienne, managing partner of a law firm in Canada, feels our software the best option to ensure compliance with discovery documentation rules."

"We believe, but are not sure, that there are about half as many law firms in Canada as there are in South America."

We're not trying to close the gaps yet (although we'll get to that), just document them. But I'm sure that doesn't feel like the whole story, does it?

CREATE AN INNOVATION BACKLOG

As you create your list of Knowledge Gaps, you may start feeling daunted again. Particularly if you aspire to Disruptive Innovation, the inventory of objects and attributes you *should* know about—but don't—probably doubled during this part of the workflow, not to mention that massive list of specific Knowledge Gaps! Once again, it's time to get organized, not panicked. You don't need to fill all these gaps immediately, but you do need to use the resources you have to fill them systematically.

As I've said before, most organizations are doing enough market research, they're just not doing enough of the *right* market research. A lot of companies don't have a plan that ties the research being done to a strategic goal or a specific question you're trying to answer. What a waste! We're going to change that with an **Innovation Backlog**.

> An Innovation Backlog details the work process of innovation by prioritizing knowledge yet to be gathered and that which is considered actionable through product, pricing, marketing, delivery, and other action. It aggregates Knowledge Statements by object and then attribute, including the confidence level and/or Score for each. The Backlog explicitly indicates whether there is a gap in the knowledge that makes it unactionable.

Consolidating everything you don't know but *want* to know about the market into a backlog is the first step to ensure your ongoing market research is geared toward innovating on purpose. Think of it as your report on and plan for innovation. It also illustrates to your cross-functional teams that you know about the blind spots in the knowledge they're working from and have a plan to fill

Create an Innovation Backlog to Keep Track of Knowledge Gaps

CREATE AN INNOVATION BACKLOG TO KEEP TRACK OF KNOWLEDGE GAPS

Product or Initiative/Innovation Strategy:

OBJECTS	ATTRIBUTES	KNOWLEDGE STATEMENT	CONFIDENCE	GAP Y/N
Governor: Regulatory Agencies	NA	"We don't know what regulatory agency sets discovery documentation rules."	Unknown	Y
Governor: Regulatory Agencies	Discovery documentation rules	"We don't know what the discovery documentation rules are."	Unknown	Y
Persona/Role: Juanita, managing partner of a law firm in Mexico	Perception of our Product	"We don't know whether Juanita, managing partner of a law firm in Mexico, feels our software the best option to ensure compliance with discovery documentation rules."	40/Believed	Y
Persona/Role: Juanita, managing partner of a law firm in Mexico.	Responsibilities	"Juanita is responsible for compliance with discovery documentation rules."	81/Recognized	N

them. Leadership can see that you're focused on finding strategic opportunities because you're investigating the market strategically and in a way that's aligned with the vision and strategy they want to achieve.

But wait—isn't a *product* backlog the same thing as an Innovation Backlog? Nope. The product backlog is an ordered list of everything you need to create a product. It's only focused on the product and only on execution of features already defined.

A product backlog is a valuable planning tool, but the Innovation Backlog—and Innovation Roadmap, which we'll discuss in the next chapter—are the tools of innovation. This is where you're documenting your quest for strategic innovation opportunities that will move your company and market position forward. It includes work being done to validate any innovation opportunity, whether it can be solved with a product enhancement or not—throughout this book I've given examples of innovating without building anything new.

The Innovation Backlog is the process of innovation; if you've followed the workflow in this book, 95% of the innovation should be done by the time you put them in a product backlog!

ACTION PLAN FOR CHAPTER 8

- Prioritize the object and attribute you know the least about and that worries you the most, and mine the Center to create Knowledge Statements.

- Establish a Confidence Language—and, optionally, a Score—to telegraph how confident you are in each Knowledge Statement. You can also aggregate the scores of each Knowledge Statement to assess confidence in your knowledge of an overall attribute of object as well.

- Create an Innovation Backlog and work with your market research, IT, and data scientists on a plan to fill the gaps in your market knowledge. Use this tool as you would a feature backlog; estimate time and resources needed, create sprints, and release the knowledge to the organization (more on that later).

Chapter 9
PRIORITIZE FILLING OF GAPS

Organize Current Knowledge

- Aggregate What You Have
- Create Knowledge Statements & ID Gaps
- Prioritize Filling of Gaps

We Market Strategists can learn a few things from our IT departments. How do they keep getting budget increases? Why do people respect their time?

One reason is because they're very, very good at communicating what they *do* with their time. From sprints to story points to roadmaps and capacity planning, they bring data and detailed forecasts to every meeting, tactical and strategic. We don't. My goal in this chapter is to help you make a plan to use your scarce market research resources wisely by gathering insight on the most important knowledge gaps first. I'll *also* help you ensure executive understanding of the value you provide for the budget dollars you spend and help you help everyone understand that your deliverables take time and you're planning to use that time to deliver the most impactful and actionable market insights to them so they can do their jobs well.

By focusing your scarce resources on gathering knowledge you currently *don't* have that you *do* need, you leverage your limited budget as best you can and maximize your ability to innovate. But now you need to show everyone how that's happening.

Once you have an Innovation Backlog, the next step is deciding which gaps to fill first. Chances are you have more gaps in knowledge than you can easily fill in the next year—and of course all the knowledge you have today could change in that time! But you have to start somewhere, and the best place is with the knowledge you need to drive your strategic goals.

Market research should always be done with an eye toward achieving strategic innovation goals. To maximize the impact of your scarce resources it should be designed to fill the most critical knowledge gaps as efficiently as possible.

Your OAS inventory now includes known and unknown objects and attributes as well as Knowledge Statements and Knowledge *Gap* Statements. Instead of expending efforts to fill those gaps haphazardly, choose to answer the questions that would have the most impact on your ability to meet strategic goals. Depending on what your strategic goals and knowledge gaps are, you'll scope your work—and prioritize your Backlog—differently.

PRIORITIZING MARKET RESEARCH TO DRIVE STRATEGIC INNOVATION

To decide what to prioritize first, return to your alignment work from Chapter 4 (By the way, are you noticing that *everything* comes down to aligning execution with strategy? I hope so). What overarching goals do you have in your contributing strategy that should guide what you learn first and what you put off until later? Some of you are tasked with opening a new market; others, with keeping current customers of current products happy and renewing. Those two directions likely result in different market knowledge needs, right?

I'm going to use Innovation Maps here in a different way than I did earlier. In this chapter, I'll use Innovation Maps to illustrate how to prioritize your market research to align with a couple of the most common organizational strategies. Once this exercise is complete, you can prioritize your Innovation Backlog to achieve your strategic goals and share that plan with an Innovation Roadmap.

INNOVATION STRATEGY: OPEN A NEW MARKET

Let's say your mandate is to expand into new markets. Your boss is looking to you for recommendations on which would be the strongest opportunity and, once identified, how best to pursue it. In this scenario, your prioritization process may look like this:

INNOVATION STRATEGY: OPEN NEW MARKETS

Market knowledge outcome desired: Find the most attractive (and accessible) market to meet your goals and gather key insights quickly.
Before you begin you must: Define key attributes of a "good market" based on strategic outcomes desired.

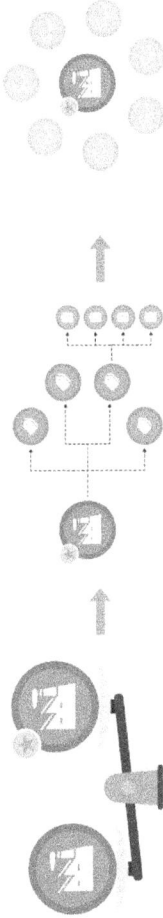

ACTION	Identify the best market to meet your strategic goals.	Research the market's ecosystem and create an O/A/S Inventory.	Gather additional, detailed insights by experimenting.
KNOWLEDGE TO GATHER	Strength of each market against the defined criteria.	Who will buy, use, and influence in this market? ID governing organizations, trusted sources of information, and alternatives.	You may prioritize finding out how they decide to buy or what regulations are changing, for example.
DESIRED OUTCOME	Team is aligned around the target market; understands why we're targeting them.	Team understands the market at a high level and can prepare experiments to learn more.	Teams focus & improve development, marketing and sales outputs with increasingly detailed market knowledge to build unleavable products.

Innovation Strategy: Open New Markets

A critical pre-work step to choosing a new market to enter is to determine what attributes a "good market" will **have**. If you want a market that's ready to buy right now, you might look immediately at the regulations that govern each market and whether one or another is required by law to have a solution like yours versus markets where solving this problem is optional.

On the other hand, if you're looking to enter a new market to increase profits, you may want to weigh the need to modify your product, hire new salespeople, or buy more advertising from market to market, to find the one with the lowest expense outlay.

Early in my career, I was asked to do exactly this. We'd launched a decision-support tool in the US that was very successful with very large companies looking to evaluate suppliers before commitment to a contract with them. My leadership wanted to globalize, and I was in the hot seat to recommend the first three countries for expansion.

I suggested what were, to me, three obvious recommendations based on my perceptions of global industry: Germany, France, and Japan. At the time, these were top industrialized countries with an emphasis on manufacturing, which seemed to make them excellent targets for our expansion. I went to the executive sponsor with my recommendation.

He said yes to France and Germany but no to Japan. When I asked why, he told me about the practice of Keiretsu.

"Keiretsu" is a Japanese term referring to a business network made up of different companies, including manufacturers, supply chain partners, distributors, and occasionally financiers. They create close relationships and work together while remaining operationally independent. Japanese companies value having close ties with one another, and the idea of working together is believed to be beneficial for all. To use software and data to choose a supplier simply wouldn't make sense to them.

I was very lucky that my executive sponsor happened also to be expert in Japanese business practices. Had that not been the case, my perception-based recommendation would've resulted in, at best, a tug-of-war with our Japanese office and, at worst, investment in an initiative that was doomed from the start.

After prioritizing a market, map the objects that play a role in the market—the ones you'll need to understand to *serve* that market. This not only includes persona functions and roles but gatekeepers, influencers, and competitors and the trends of the market itself. Some of this may have already come out in earlier steps, but a quick brainstorm now is a good idea. With objects identified, you can list attributes to investigate for each and the sources from which to get that knowledge.

Finally, prioritize the need for knowledge of each of the objects, and make a plan to gather it.

You might be thinking, *"This a lot of cycles—how can this be more efficient than what we're already doing?"* But imagine you're faced with globalizing a product. Where do you start? How do you decide? It's better to spend a *short* period of time in structured planning versus endless hours in meetings guessing and advocating and more time after that rethinking your guesses and cajoling people to get with a program they don't agree with. Further, you don't have to wait for all the research to be done before acting—you shouldn't, as a matter of fact. Once you've chosen the market, you should put positioning, pricing, and product experiments out to test hypotheses and find what will work. More on that later.

Your organization may have other strategic goals. Here are a couple of examples of other innovation strategies, and how each strategy determines what you need to know about your market:

INNOVATION STRATEGY: BE #1 IN A MARKET

Your contributing vision and strategy might be more focused on dominating the competition than entering new markets. You could deliver that objective by identifying the alternatives your market is choosing from and then learn everything you can about the current dominant alternative. Note that "competitor" could also mean alternatives other than buying a product.

Innovation Strategy: Competitive Primacy

INNOVATION STRATEGY: COMPETITIVE PRIMACY

Market knowledge outcome desired: Be the #1 provider for a market.
Before you begin you must: Align around the market in which you want to achieve dominance.

ACTION	Win/Loss analysis: ID competitors	Win/Loss analysis: Document buyers and buyer journey	Win/Loss analysis: Why we win/lose
KNOWLEDGE TO GATHER	What are the alternatives considered by this market? Include indirect competitors such as workarounds and substitutes.	Who will buy and influence in this market? ID governing entities and trusted information sources. Document the buyer journey.	Why are you winning? Why are you losing? Include factors beyond the product such as pricing models, reputation, interoperability, etc.
DESIRED OUTCOME	Set the baseline to systematically neutralize all competitive threats.	Understand the buyer journey & what Objects are involved in it, in order to address roadblocks to winning in the market.	Focused development, marketing and sales outputs will increasingly build unleavable products and win more deals.

INNOVATION STRATEGY: DISRUPTIVE INNOVATION

And what about that elusive, exciting goal—finding the next new thing: a big idea that results in you changing the way a market solves its problems or achieves its own success.

Here, it's critical to make sure you and your leadership agree that this is an activity that you want to pursue. Disruptive Innovation can be a high-reward strategy, but it's also high risk, expensive, and time consuming. Ideally, you've already aligned on this (and they've agreed to fund the effort) back in Chapter 4.

With that alignment, choose a market and decide what types of problems you want to solve for them at a very high level. Without this step, you'd have to investigate everything in the world, all at once, all the time. Some scope is necessary! Instead of saying, "We are looking to do more for law firms!" dial in a little more. For example, "Let's watch for changes in the way German corporations interact with their outside legal counsel, to see if we can help those law firms attract more clients and expand current client relationships."

Will this kind of focus cause you to miss a Big Thing you could've capitalized on? Yes. Will you be able to innovate on purpose without it? No.

Once you've defined scope, identify sources of knowledge by doing an extended OAS inventory. Be sure to include market workflows, regulations, and influencers as well as the markets and personas themselves. Plug into these sources and begin to aggregate data.

Be sure to keep your horizons broad. You're looking for tomorrow's innovation opportunities, buried in what you're learning today. Synthesize your data to review trends, shifts, and changes that, taken together, may create exactly the opportunity you're looking for.

Innovation Strategy: Disruptive Innovation

INNOVATION STRATEGY: DISRUPTIVE INNOVATION

Market knowledge outcome desired: Find the next new thing.
Before you begin you must: Ensure leadership will fund this effort. Manage expectations. Have a clear understanding of organization vision.

ACTION

Choose a scope that aligns with organizational vision.

Gain internal alignment around what market, set of market needs, trends, workflow, technology, or other focusing elements you'll use to map the effort.

Create an extensive O/A/S inventory.

Create a plan and method to aggregate knowledge sources for evaluation.

Identify trends and synergies within the ecosystem. Investigate solution enablers that may or may not be currently used by your focus market.

Identify opportunities to solve an emerging market problem or solve an existing problem in a radically better way.

KNOWLEDGE TO GATHER

Critical focus for a broad effort.

What sources of insight should you be watching for signals of a disruptive opportunity.

DESIRED OUTCOME

A network of data sources and plan to monitor, synthesize, and evaluate them.

PRIORITIZE YOUR INNOVATION BACKLOG WITH SCORING

With this knowledge, you can prioritize efforts to fill the knowledge gaps detailed in your Innovation Backlog.

A way to prioritize is to use a simple scoring system. Your scoring system should be built to deliver priorities that are meaningful to you. In the graphic below, I've used impact and urgency, each on a scale of 1–5. Impact reflects the significance of knowing this information on your future innovation activities. Urgency signals if your need for this information is immediate or can wait. The priority score is the product of the impact and urgency scores. You should create a system that works for you and be sure to add definitions and unifying examples for each score level.

Earlier, I gave some examples of Confidence Language and Innovation Maps that could be used with cross-functional and leadership teams to illustrate your progress against priorities. Another good tool for that is an Innovation Roadmap.

CREATE AN INNOVATION ROADMAP

Most organizations have a roadmap—maybe several—showing how *product* builds will progress. Sometimes they're even explicit about the buyers and users those products will serve and the corporate strategy they fulfill. But what about the work that should be happening *before* a product build begins? While most companies roadmap what they'll *build*, no one shares a plan for what they'll learn *in order to* build. We're going to change that.

The **Innovation Roadmap** represents the milestones to fill gaps in your market knowledge and identify opportunities to innovate. Like any roadmap, the point is to connect your market research activities with the strategic goals they're intended to drive. This connects to the point I made in the last chapter, that product roadmaps and backlogs are planning tools, but the Innovation Roadmap and Backlog document the process of innovation.

An Innovation Roadmap expresses high-level knowledge gap summaries from your Innovation Roadmap and shows when you plan to fill them. It connects your market research activities to the strategic goals they're intended to drive.

While I've included an Innovation Roadmap example, yours can look different and include additional elements such as type of research you're going to conduct and resources used for each research initiative. However you design your Innovation Roadmap, be sure to include the following elements:

- *Corporate and contributing vision and strategy.* This reminds your audience of the big picture, and how you fit into it. It also keeps you honest. As you start to fill in your roadmap, you can look at the top of your page to see whether you're following your stated focus or doing a lot of market research you don't need to do.

- *Objects and attributes of your research.* Give the team a sense for who you'll be learning about and what you want to learn. Not only does this share good information, it inspires your cross-functional team to keep a special lookout for insights on the attributes and objects you're especially concerned with at the time. Instead of the daunting, "Bring me anything you hear," teams understand, "Hey, I should be listening to my Canadian prospects for any regulations they have to follow and figure out who's in the buying workflow."

- *Timeline.* This is really helpful if you share research resources with other teams. Ideally, you'll get together ahead of time to make sure you aren't dogpiling your data science team in Q2. Even if pileups are unavoidable, at least you'll all know when they'll happen.

PRIORITIZE YOUR INNOVATION BACKLOG

Product or Initiative/Innovation Strategy:

OBJECTS	ATTRIBUTES	KNOWLEDGE STATEMENT	CONFIDENCE	GAP Y/N	IMPACT	URGENCY	PRIORITY SCORE
Governor: Regulatory Agencies	NA	"We don't know what regulatory agency sets discovery documentation rules."	Unknown	Y	5	5	25
Governor: Regulatory Agencies	Discovery documentation rules	"We don't know what the discovery documentation rules are."	Unknown	Y	5	4	20
Persona/Role: Juanita, managing partner of a law firm in Mexico	Perception of our Product	"We don't know whether Juanita, managing partner of a law firm in Mexico, feels our software the best option to ensure compliance with discovery documentation rules."	Believed	Y	5	3	15

Prioritize Your Innovation Backlog

Innovation Roadmap: A Plan to Learn Before You Build

INNOVATION ROADMAP: A PLAN TO LEARN BEFORE YOU BUILD

Corporate Vision & Strategy Expand product line to offer end-to-end workflow solutions, first for law firm Finance departments, and then for all aspects of running a law firm. Enter the Canadian and U.S. markets.

Contributing Strategy Open the Canadian market with first sale in Q2. Revenue of $30M by 20XX. Maintain South American renewals. Enable success of new product offerings by ensuring interoperability.

OBJECTS	ATTRIBUTES	Q1	Q2	Q3	Q4	1H	2H	YR3
Vivienne: Managing Partner (CAN)	Buying Workflow Unique Needs			████				
Adela: A/R clerk (MEX)	Software Priorities Workflow Changes				██		██	
Governors: (CAN)	ID CAN Governors Governing Rules	██	████					
Our LegalcoEG Team	Product Roadmaps Interoperability Reqs						█	█

The Innovation Roadmap is a great tool to share with the rest of the organization so they understand what you will—and won't—be discovering about the market in the coming months.

At this stage you should be using the knowledge you've already aggregated to set priorities and make decisions, but there's always more to learn. With gaps identified, what's next?

ACTION PLAN FOR CHAPTER 9

- Review your contributing strategy and Innovation Map. Based on your strategic goals, what Knowledge Gaps should you fill first?

- Create a Backlog scoring system using the knowledge above. Score your Backlog to prioritize what insights to pursue first.

- Summarize this effort in an Innovation Roadmap.

- Prepare to share the difference between product and Innovation Roadmaps and Backlogs with your leadership and teams.

Chapter 10
INVESTIGATE

Find Opportunities

- Investigate
- Experiment & Validate
- Interpret

It's time to turn our attention from organizing the knowledge you already have to finding new opportunities. This short chapter will make that transition by exploring the difference between data and knowledge and laying the groundwork for investigating in the market.

According to Wikipedia, a hypothesis is a proposed explanation for a phenomenon. For a hypothesis to be a *scientific* hypothesis, the scientific method requires that one **can** test it and demands that you **do** test it. Scientists generally base scientific hypotheses on previous observations that can't be satisfactorily explained with the available scientific theories.

In the world of the Market Strategist, a hypothesis is something you believe to be true about your market but don't have enough information to treat is as a Recognized piece of knowledge. It could also be a response (product or otherwise) that you believe will fix a weakness or serve your market that's been inspired by the data you have. Hypotheses aren't the only tool you'll use to discover innovation opportunities, but I'm breaking it out here to help make the point about data versus knowledge.

DATA IS ONLY THE BEGINNING

Data and knowledge are different. It's the difference between knowing there's an **opportunity** to act and the **ability** to act in a way that solves your market's problem and enables your success. To innovate on purpose, it's critical to know the difference between the two and construct your innovation workflow to find both.

Data are often the first market and organizational indicators you gather, giving a signal that something's very right or very wrong. Ticks up in NPS or other continuously gathered scores are a good example. Later in your innovation cycle, data quantifies "how many," "how often," and "how high a priority." Data is usually gathered in surveys, interviews, and analyst statistics. Data's great; you can't have knowledge without data. That said, a single data point on its own is seldom actionable.

My experience has shown that organizations frequently jump from data right to a solution, likely skipping over a critical hypothesis that should be tested. This is one of the single biggest time-wasting workflows most organizations embrace. You know how it goes:

"Hey—our customer service NPS scores are down three percentage points from last quarter. **Do** something!"

"Right—we'll implement a new chatbot to improve user experience!"

"While you're at it, put a new KPI in place for customer service calls: every call must be resolved within two minutes!"

"Great! And, for good measure, we'll sign all the CSRs up for a round of training!"

Six months later . . .

"Scores are down again—**dooooo** something!"

"Right. Now we're going to . . ."

What if, instead of moving from "data" to "solution," you add an investigation step in between? Something to test a response or learn some specifics before you roll out a lot of "responses" that don't, actually, solve the problem. Here are three investigation tools to consider: questions, hypotheses, and experiments.

QUESTIONS, HYPOTHESES, AND EXPERIMENTS

You may think of others, but questions, hypotheses, and experiments are key tools to turn data into market-driven innovation opportunities. You won't use all of them for every data point all the time, but each should be considered.

Questions: are what you need to know to respond to a problem that may have been uncovered by statistics or scores. What do we need to know to respond? For the above NPS score example, I can think of quite a few questions we might want to ask:

- *Market segmentation question.* Was there a **specific persona or market segment** that scored us negatively? Knowing what part of the market is unhappy may or may not lead you to the right response but will certainly help you pinpoint the problem.

- *Product segmentation question. **What part** of the customer service experience did they find to be poor? Scores are valuable flags to indicate trouble, but a poor score may not mean the entire offering or experience was poor. It's inefficient and risky to "fix" an entire product when a quick survey could tell you the specific element that needs tweaking. If you guess, you could even mess up the good parts of the product without realizing it!

- *Causal question.* What's **causing** the experience to be poor? Another good question. A causal question may get to problems beyond the product. For example, maybe customer service is bad because I have to call you, or your chatbot isn't staffed 24/7. In other words, it's not the customer service that's bad, it's the delivery mechanism.

- *Resolution question. What does "great performance" look like* to the group that's having the problem? Perhaps this group has higher expectations of customer service that others, or perhaps their preferred solution is different. Using the previous example, maybe only younger customers want to use your chatbot at all, or maybe Canadian customers prefer email while those in New Zealand prefer texting.

Asking questions is critical! Your chatbot implementation won't make a bit of difference if you need to overcome a language barrier or your return/exchange policies are terrible. It also won't matter if people are happy enough with your chatbot but hate your phone service. And overcompensating when the market would be happy with a simple change just means you squandered scarce resources that could have been used elsewhere.

When my husband Tom and I first moved to Austin, we read some fairly heinous reviews of our local hospital. Most of the reviews were three out of five stars—not exactly stellar when you're putting your life in their hands! But when I read a few of the reviews—and then a few more—what I discovered was that the community felt that the hospital care, staff, facilities were all top-notch. On the other hand, the billing department was a mess!

That's not a great thing to discover about your local hospital, but it's not like they were accidentally removing the wrong organ or dropping Junior Mints into people. My point? If you're going to solve a problem, you need to know what the problem *is*. Let's keep building the bridge between data and knowledge with hypotheses:

Hypotheses: are actionable suppositions (aka guesses) based on whatever data you have that you can test to validate a course of action. You often brainstorm hypotheses as a result of questioning data. One piece of data could deliver several hypotheses, each of which could lead you in a unique direction. Let's use the same examples from the Questions section above:

- *Market segmentation hypothesis.* "We think the lower scores are coming primarily from our customers in Mexico."
- *Product segmentation question.* "We think it's the language barrier that's causing dissatisfaction with our customer service: our CSRs don't speak Spanish."

- *Causal question.* "We think our customers in Mexico feel unappreciated when they can't speak their own language with a CSR."
- *Resolution question.* "We think having Spanish-speaking CSRs available for all Mexico customers would solve the problem."

We use hypotheses for the same reason a scientist does: as a tool to articulate the answer we believe *could* be true, as a precursor to determining whether it *is* true.

Experiments: are what you use to validate/discredit your hypotheses. How could you know whether your hypothesis is true? Here, it could be through further analysis of the data, surveys, or focus groups.

Questions, hypotheses, and experiments are the building blocks needed to transform data into knowledge. But isn't all this questioning and hypothesizing a huge waste of time, classic "analysis paralysis"? Not if you're doing it right.

True, you will spend more time talking to the market. But my real answer to that question is: What's more cost- and resource-efficient, guessing by building a product and hoping the market will buy it, or doing a couple of A/B tests and a focus group?

And as I've said before, the idea isn't to wait until you have perfect market knowledge or 100% confidence that your innovation will succeed—because you never will—but to continue to close the Knowledge Gap to innovate more and more successfully over time.

We'll use questions, hypotheses, and experiments to fill the Knowledge Gaps you have in your Innovation workflow.

ACTION PLAN FOR CHAPTER 10
- Identify a current "red flag" score or data point that concerns you. Create a list of questions, hypotheses, and experiments you could use to transform the data into actionable insight.

Chapter 11
EXPERIMENT & VALIDATE

Find Opportunities

- Investigate
- Experiment & Validate
- Interpret

In my experience, most organizations are doing a lot of market research. They're mining data, leveraging crowdsourcing techniques, doing tons of surveys, and scheduling a focus group each week. As a result, they've got tons of data—but not much insight.

Sometimes it's because they confuse data with knowledge, sometimes because it's expensive and difficult to reach the right people.

I avoid the term "market research" as much as possible because market research *is* innovation—and vice versa. You can't get the latter without the former except through blind luck. It's during experimentation and validation that you really kick off the innovation process, but it *is* expensing and time consuming. Therefore, it's critical that you continue with the focus and prioritization discipline you've put into the workflow so far, to keep this work both highly valuable and *cost* effective.

In this chapter I'll offer some tips for asking the right questions to get actionable innovation ideas from your markets.

TIMING IS EVERYTHING

Where are you in your innovation process? Are you just learning about your market, trying to fill major knowledge gaps? Or are you considering improvements to an existing and much-loved product due to suspected preference changes or rumors of a new competitor?

Purpose-driven innovation is enabled through constant conversation with your market. If your conversation is ongoing, thorough, and systematic, you're never waiting for new data to decide what to build and how to sell it, but you're always learning something that will compel you to hone and improve. A continuous cycle of hypotheses and validation is how you break this conversation into usable chunks that benefit your market while also fulfilling your innovation strategy.

While this is an ongoing and often cyclical process, it helps to think of it as a progression. At the highest level, your goal is to move your overall knowledge from Believed to Recognized.

GETTING FROM BELIEVED TO RECOGNIZED

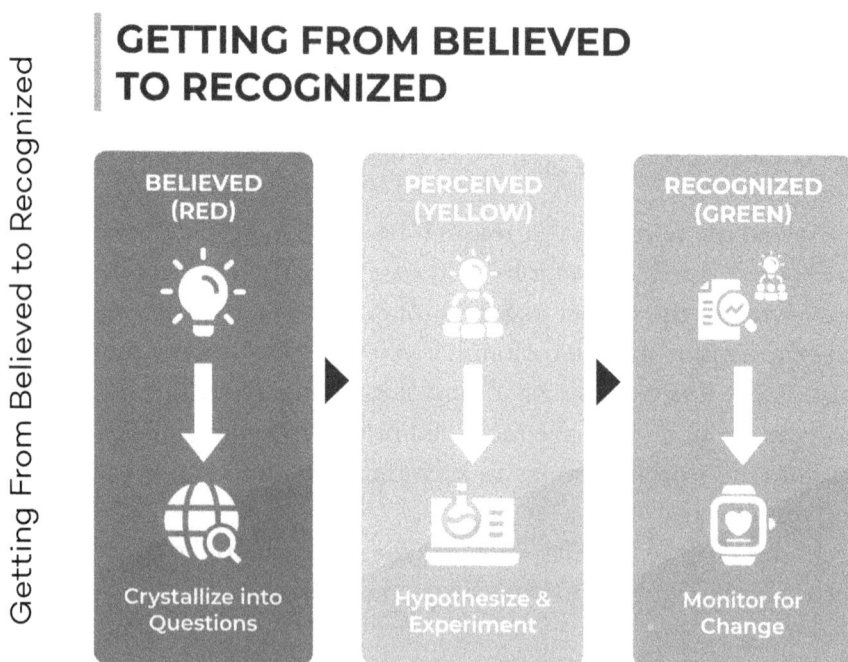

BELIEVED (RED)	PERCEIVED (YELLOW)	RECOGNIZED (GREEN)
Crystallize into Questions	Hypothesize & Experiment	Monitor for Change

When you believe something to be true about your market but don't know, the next step is to craft questions and/or hypotheses that will **help** you know. If you have indicators that this belief might be true, it's time to put experiments or validation questions into the market to be sure and to get details you can respond to. Once you have enough information to meet your organization's "Recognized" threshold, it's time to put monitoring in place in the event the knowledge changes—more on that in Chapter 14. Now, let's break this conversation down into more detailed phases:

HAVE A CONTINUOUS CONVERSATION WITH YOUR MARKET

Innovation on purpose requires a continuous conversation with your market. We've already discussed how to know where to focus your efforts, but I want to remind you of that here, before we dive into the details of timing.

OUR GOALS?

Your organization has a corporate vision of who you want to be and a high-level strategy for how to get there. Earlier, you tightened that overall focus into your own contributing strategy that focuses your activities on those that contribute to corporate goals. You're not listening to every object in every market all the time, but rather those objects that deliver inspiration that helps you down the path of who you want to be.

That said, the more extreme your innovation goals are, the deeper your object inventory and the longer your horizon will be. For Reactive Innovation, your object inventory is restricted to your existing customers and the time horizon, a quarter or two. For Disruptive Innovation, you will be investigating a broad ecosystem of objects on a horizon that could span many years.

Whatever your organizational vision and innovation goals, this understanding provides focus for the rest of the conversation.

WHAT'S NEW?

Something new and unknown can be risky and even a little scary, but it should ignite innovative thinking. "What's new" is a logical starting point in any market. Often, this is how new businesses—new industries—are born; somebody sees a new need, or a new ability to solve a problem, and decides to go for it. Often, the answer to "What's new" isn't going to come from asking a question but rather observing, identifying gaps, or synthesizing market clues to form a broad hypothesis. This phase is about finding and naming an emerging need or desire.

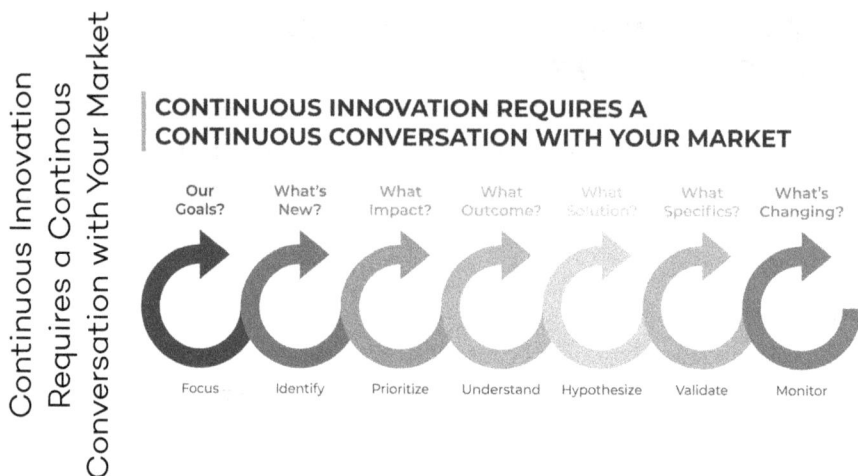

CONTINUOUS INNOVATION REQUIRES A CONTINUOUS CONVERSATION WITH YOUR MARKET

WHAT IMPACT?

Not all problems or desires are equal in the mind of your market. When you question the market about impact, you're investigating the consequences or challenges that result from this unmet need or desire, in order to prioritize your own next steps.

WHAT OUTCOME?

The fourth element in a continuous conversation is asking what outcome your market is trying to achieve. What would "solved" look like to them? Your goal is to understand what they want, not brainstorm what you're going to build them—not yet.

WHAT SOLUTION?

Almost a cycle within a cycle, the discussion of solutions and specifics takes place in the market but also inside your own company, and this is where you create hypotheses for the market to react to. Whether it's marketing tools, products, pricing, or contract terms, this discussion is about what solution makes sense for the market. With more complex solutions, you may need a second step.

WHAT SPECIFICS?

If your solution or the outcome desired isn't straightforward, you'll likely cycle through hypotheses and responses as you get closer and closer to what the market wants. Contrary to the startup legends, most organizations that deliver innovative solutions to market go through more than one test to do it.

WHAT'S CHANGING?

Too many organizations stop listening to the market after they launch version 1.0 of their solution. The next iteration is often on our timeline, not the market's, and forever focuses the roadmap on iterative tweaks requested by existing users. In other words, we're at risk of being disrupted almost as soon as we enter the market. That's because needs change, even for workflows that've been around for decades. The tools available to solve them change, the people doing them change, and getting caught up in continuous improvement—without looking for the larger changes happening in the market—could result in a customer base that loves you, right up to the point where they leave you for the innovative start-up.

Establish questions and methods to monitor the market for evolution—and revolution—once you've entered it. Likely, this means keeping in touch with the *managers* of your users, the governors and gatekeepers and enablers too.

THE CYCLE OF CONTINUOUS CONVERSATION

How long you spend in any phase is dependent on factors unique to your business, such as your innovation strategy, market needs, and types of offerings you build. But whatever the specifics, your goal is not to spend a certain amount of time or effort asking questions in each phase but rather to get what you need as quickly as possible to progress to the next.

Have a Continuous Conversation with Your Market

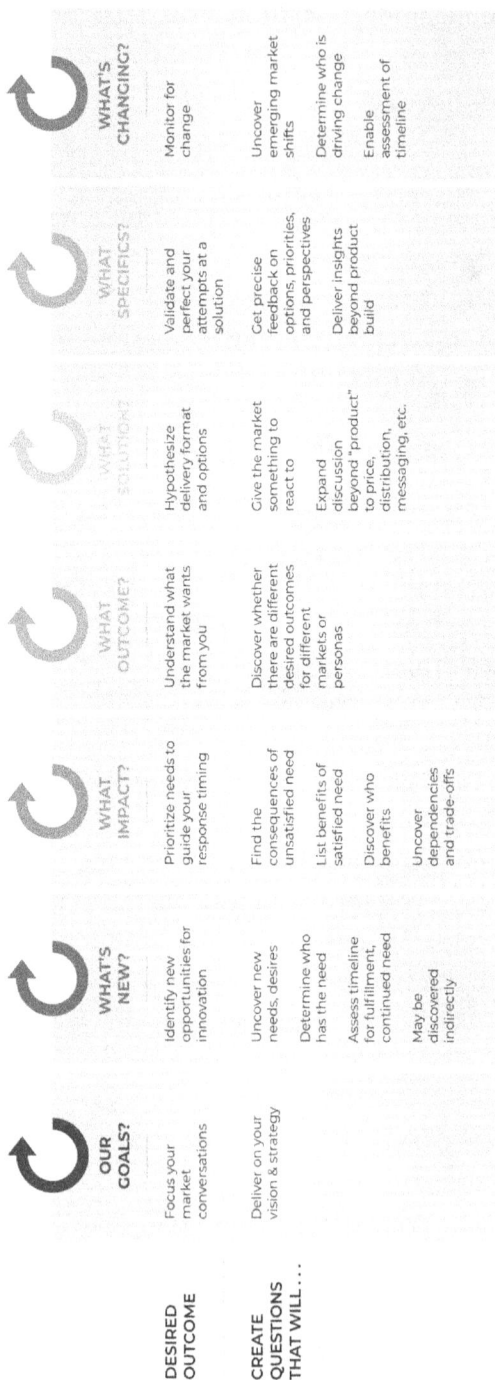

HAVE A CONTINUOUS CONVERSATION WITH YOUR MARKET

	OUR GOALS?	WHAT'S NEW?	WHAT IMPACT?	WHAT OUTCOME?	WHAT SOLUTION?	WHAT SPECIFICS?	WHAT'S CHANGING?
DESIRED OUTCOME	Focus your market conversations	Identify new opportunities for innovation	Prioritize needs to guide your response timing	Understand what the market wants from you	Hypothesize delivery format and options	Validate and perfect your attempts at a solution	Monitor for change
CREATE QUESTIONS THAT WILL...	Deliver on your vision & strategy	Uncover new needs, desires Determine who has the need Assess timeline for fulfillment, continued need May be discovered indirectly	Find the consequences of unsatisfied need List benefits of satisfied need Discover who benefits Uncover dependencies and trade-offs	Discover whether there are different desired outcomes for different markets or personas	Give the market something to react to Expand discussion beyond "product" to price, distribution, messaging, etc.	Get precise feedback on options, priorities, and perspectives Deliver insights beyond product build	Uncover emerging market shifts Determine who is driving change Enable assessment of timeline

ASK THE RIGHT QUESTIONS

There isn't a list of "magic questions" that will lead you straight to a Disruptive Innovation opportunity, but I will give you some tools to help you deliver actionable insights for whatever innovation phase you aspire to.

LEVERAGE CYCLES OF VALIDATION

Action always seems better than study. "Do something!" is the rallying cry of business, and who wants to sit by while a product's revenue tanks or an opportunity is picked up by the competition?

But the jump to action—whether it's fixing a problem or introducing disruptive innovation—generally results in more time and resources spent and less success in the market. Better to create a constant conversation with your market driven by your priorities and their biggest needs and wants that allows you to home in on specifics and deliver increasingly more resonant and powerful solutions.

Questions will be different at each phase of the conversation—the graphic below gives you one example of how that might play out:

Immediately putting a few likely solutions into the market, as seen in the "Typical" sequence, seems much more responsive than the second "Better" option. But think about the time you'd expend convincing leadership, marcomm, and sales that each of these was a good thing to do. Think of the time and budget you'd use to create all those options. And without putting test parameters in place, how would you know whether they were working?

Get to market with a solution as quickly as possible, but don't build when you haven't gathered enough information to do so. In the "Better" sequence, the two additional questions are very quickly answered, perhaps with just a second review of your data analysis and a one-question survey to the customers in Virginia and Georgia. With these quick cycles, you'll build tools and workflows that are efficient tests and more likely to succeed, while also saving time and building credibility along the way.

One way to implement the conversation method of innovation is to keep the conversation going with what I call "extension questions."

Market Validation—Right and Wrong

MARKET VALIDATION - RIGHT AND WRONG

"Quick action" can be more wasteful and less successful than quick cycles of targeted learning

TYPICAL VALIDATION/ACTION SEQUENCE

TIME/EXPENSE/REWARD
High/High/Low

Identified = "Southeast customers are less likely to renew than Midwest customers."

➤

Action = try everything: "Have sales offer a pre-renewal discount in the Southeast and do an email campaign and a webinar with best practices and..."

BETTER VALIDATION/ACTION SEQUENCE

TIME/EXPENSE/REWARD
High/High/Low

Identified = "Southeast customers are less likely to renew than other customers."

➤ Specific states or sales territories?

➤ 1st year or all?

➤ Numbers & revenue impacted?

➤ Average number & renewal?

➤ Why?

Prioritize = "Year-1 renewals in VA & CA are 65% vs 90% nationwide. Impact is $10M in lost revenue."

➤ Why?

Understand = "Top reason is that users never start using the product."

➤ Why?

Hypothesize/Validate = "Will building awareness of our onboarding toolkit help?"

➤

Action = test solutions to specific problems: "Provide Sales Ops with welcome tools and create an email welcome and 10-day follow-up; monitor results."

ASK EXTENSION QUESTIONS

Extension questions are designed to take you to the next level of knowledge as efficiently as possible. You can also think of them as follow-up questions to hypotheses or something interesting you hear in a broad interview or find during analysis.

You'll come up with dozens of extension questions, but here are a few to get you started:

- *What else?* Product managers, developers, and UX professionals are excellent at their jobs, but their focus is usually to create a product to solve a problem. Often these individuals aren't responsible for pricing, marketing, selling, or positioning their solutions, so they don't listen for market insights on these topics. I've discovered more than one team that was **intentionally disregarding** these insights because they were "out of scope." This results in a massive opportunity gap. Even reactive innovators often miss definite market signals for new pricing models, different distribution strategies, or easier implementation because it's "not about the product." Don't make this mistake! We'll talk more about this as "360 Degree Innovation" later.

- *When?* Knowing the market's timeline can help you prioritize your own. Ask whether they have the need right now or see the need coming soon. Another great line of questioning is whether they see an *end* to this need. Again, if so, when? Finally, what drives the timing of the need? Is there a trigger versus an actual date?

  Several years ago, I worked with a data scientist who analyzed what had triggered our new clients to buy our products (workflow-enablement software). At the time, we were marketing to all small law firms, whether they had one attorney or 15. What we discovered was that, until a law firm employed five attorneys—not people, but attorneys—*they didn't buy*. Collaboration tools just weren't necessary at this size—especially when budgets were still tight and the priority was building the practice. When they reached the five-attorney mark, they became very active prospects for our products. It was a trigger we were able to isolate to become much more effective with our lead generation.

- **Priorities.** Personas will be happy to tell you what they want: what they'd love to have or really should have. But how important is this thing to them? Is one item more important than another, and are some items so important any solution would be worthless without it? One of our key goals in innovating on purpose is to prioritize the innovation opportunities we go after. Asking about relative importance and "table stakes" is a great start.

- **Predictions.** You're not just innovating for today but also to build loyal customers in the future. Getting an idea of what they think is **going to** happen is often a good way to direct your discussions and plan ahead. A good follow-up is "When do you think this will happen?"

All of that said, you should remember that much of the market validation you do may not be through interviews with personas, or anyone. You may not be "asking" anybody anything yet still learn a lot.

Now that you know **what** to ask, you need to determine **who**.

ASK THE RIGHT PEOPLE

If you ask a five-year-old if they want a candy bar, the answer is likely to be yes. But is it the child's decision whether to have one? Probably not. That's a classic example of a user versus a buyer, and you may already be asking questions of these folks. But what if you're trying to figure out what kind of candy bar a five-year-old is going to want in three years? What if you're trying to figure out what kind of candy bar the parents of the average five-year-old will buy in three years? Figuring out who to ask innovation questions can be challenging without a plan.

Depending on what you're trying to learn, it's possible to ask the right question of the wrong person and get an answer that sounds logical, actionable, and doable but may lead you to offer the wrong solution, an incomplete solution, or a solution that no one will spend money to buy.

We covered some of this section in Chapter 6, when I outlined objects, attributes, and sources. Using those definitions, I'll expand on some additional considerations for finding the right person to ensure you get the information you need.

QUESTION THE CORRECT OBJECT BASED ON THE ATTRIBUTES THEY EXPERIENCE OR CONTROL

Product users are the right ones to speak to when it comes to product features they need fixed and new features they want added. *Managers* of your users can tell you what those users will need next year. Buyers can tell you how they make the buying decision. But who trains new hires on the product? Who hooks it up to the other systems it needs to interoperate with? Are there users with different priorities, demographics, and workflow preferences? To get an answer that truly guides you and minimizes your risk of failure, get to those who actually experience the challenges.

What if you can't? More on that in a minute. You may need to start by asking current product users questions simply because you haven't figured out how to get to anyone else. If so, remember that you're asking them about how *someone else* thinks, behaves, and makes decisions. Be careful how you word questions with these folks!

QUESTION THE CORRECT OBJECT BASED ON YOUR INNOVATION PHASE

The more disruptive your innovation goals, the longer-term perspective your source must have. If you want to know what a company has been waiting for you to build for the past year, ask a user. If you want to know what they'll need three years from now, ask the CEO or CTO.

QUESTION BEYOND THE USERS AND BUYERS

Chances are, there are a lot of insights you could gain from experts inside your company (Chapter 5's Contributing Teams) and objects adjacent to your market (influencers, gatekeepers, governors) that you're likely not leveraging. I listed those earlier but want to remind you. Rope in your IT, finance, and legal departments to keep you up to date on the latest technical capabilities, buying behaviors, and pricing options and new regulations. Interview the officers of industry associations or regulatory bodies to get updates on what's happening in your persona's world.

- *Who else?* We often think in terms of buyers and users, but what about administrators, operations teams, procurement? Sometimes we end up blocked from an easy sale because we made it too hard to buy or implement our terrific product.

- *Perceptions.* Ever tried to interview a CFO? The managing partner of a law firm? Sometimes you can't get to the person whose insights you need to innovate on purpose. I want you to try, but what if it's impossible? In that scenario, ask those who you *can* get access to to give you their perspectives on how your real object will behave when making a decision. With this technique you're at least getting insight into what those others believe your object may do, versus their own personal opinion.

REMEMBER TO OBSERVE

Particularly when your goal is Disruptive Innovation and you're actively looking for inspiration, that next new thing, your market may not have an answer. They may be living with a problem they don't think about anymore, assuming there is no solution. Or they may not be able to visualize the next new thing that they would love to have, because they don't have the technical or other expertise necessary to build it in their minds. How do you ask the market when the market doesn't know what to ask for?

Instead of surveys and focus groups, you may be observing behaviors of personas, influencers, or governors in the market. What attitude to they take toward each other? What frustrates them?

Early in my career as a product manager for B2B workflow enablement software, we were preparing to do some onsite observation of users of our product. My manager told us to look at the sticky notes on the edges of their computer monitors. "Anything that's not a childcare phone number, pet picture, or cartoon is an opportunity for us to improve our product" he said, "make a note and ask about anything you see that looks like a reminder, password, or other gap in our product." He was right—we came back with two clear improvements to the user experience that we would never have thought to create.

GETTING PEOPLE TO TALK TO YOU

When I was instructing full-time with Pragmatic Institute, one of the biggest challenges I heard from product managers and marketers was, "I can't get anyone to talk to me!" My teams faced this too. How do you get busy people—especially executives or someone who will never buy your product—to talk to you? Following are a few tips I've picked up over the years:

- *It's a numbers game.* The reality is, you'll have to reach out to a lot of people who won't even acknowledge you. My informal polls say that, for tech companies, you'll have to reach out to 5–25 people to get one interview with an existing customer, and 25–75 for every non-customer interview! Don't get daunted.

- *Be opportunistic.* Tuck questions into every interaction you have with the market. People who won't agree to an interview will answer one question at a time over a six-month period as they make a buying decision or get a call from you.

- *Make it easy to help you.* Make sure everyone who touches the market has questions they can ask and an *easy* way to get you that they hear. Most of your team members probably hate filling out forms—if they can remember where the form is. Set up Slack or Teams channels, email, voicemail, and any other collaboration tools your company uses so teams can forward you insights in whatever way is most convenient for them.

- *Ask questions that intrigue your object.* You need specific insights, but buyers need questions they can sink their teeth into! What do we all like to talk about? Ask for predictions, opinions, frustrations—topics that get almost anyone to speak up.

- *Keep it short.* The second-best tip I've heard is to ask for a 10-minute meeting. I've had many folks tell me that, if they respect the 10-minute time, their interviewee will extend the interview 50% of the time.

- *Never underestimate the power of a gift card.* It doesn't matter if your buyers are executives. It doesn't matter if they're scientists, scholars, or other serious types. For years I've asked product teams whether they've had success with gift cards, and the answer has been a resounding "yes!" If buyers aren't barred by law or regulation from taking a gift card, they love them! So break out the piggy bank, and pay for a bit of someone's time. Use values that work in your world, but as a suggestion offer anyone who will do even a short interview or survey US$50. An hour or focus group? $100–500.

- *Give to get.* Often, the information you gather is interesting to the subjects, too. If so, and if you can maintain individual privacy—share!

USE THE RIGHT METHOD: TOOLS, MEDIA, AND VENUE

Earlier I mentioned the need to be opportunistic when asking questions. Keeping it easy for your source is important too. To achieve this, take a moment to align the tools, media, and venue to ensure you get the most and best market insight possible for your efforts. Choosing the right method could also mean enabling many methods of gathering information.

For those of you with a market research or data science team, chalk up another benefit in your quest to innovate on purpose. If you have a team that's trained to use the right tools, you'll get better market insights faster and more efficiently. But what if you don't? Then you need to answer one more thing, which is how will you ask this question?

The good news is, now that you know the "what" and the "who," the "how" should be easier to figure out. Think of three points to clarify:

TOOL

By this I mean the method you'll use to ask the question. Can you use a survey, or does this question lend itself to an interview? Should the interview be one-on-one to respect confidentiality, or would you get better insights in a group/brainstorming environment? Maybe you won't even **ask** but rather observe or study.

MEDIA

Would this conversation be better held face-to-face, or can you do it online? In some areas of the world an email or text is considered overly informal, perhaps even rude. In others, you'll never get your subject to answer the phone, so email is best.

VENUE

Where will you administer the tool? Can you post a survey on Facebook or LinkedIn? Should you have a meeting in their office or a focus group site? While venue is often driven by media, you should spare a moment or two to think through what's most likely to work for your subject and get you what you need.

Experiment & Validate

EXPERIMENT & VALIDATE

Product or Initiative/Innovation Strategy:

OBJECTS	ATTRIBUTES	KNOWLEDGE STATEMENT	PRIORITY SCORE	WHAT IS THE QUESTION?	WHO SHOULD I ASK?	HOW SHOULD I ASK?		
						TOOL	MEDIA	VENUE
Governor: Regulatory Agencies	NA	"We don't know what regulatory agency sets discovery documentation rules."	25	What is Juanita's 3-year strategic plan to improve the revenues of her law firm	Juanita, Managing Director of a Law Firm in Mexico	Interview	Phone request from our VP/ Zoom or in person interview	Juanita's office or lunch meeting.
Governor: Regulatory Agencies	Discovery documentation rules	"We don't know what the discovery documentation rules are."	20	Do lawyers in Mexico feel our tool is the best option on the market?	Juan, legal associate	Spanish-language survey	InMail solicitation	LinkedIn
Persona/Role: Juanita, managing partner of a law firm in Mexico	Perception of our Product	"We don't know whether Juanita, managing partner of a law firm in Mexico, feels our software the best option to ensure compliance with discovery documentation rules."	15	What are the top challenges lawyers in Mexico face when using our tool?	Juan, legal associate	Workflow Observation	Email request from our head of training.	In-Office

AVOIDING ANALYSIS PARALYSIS

You were probably worried about analysis paralysis seven chapters ago, right? Isn't this too much? Will we ever build anything again? I hope, if you did start worrying about this, that you saw the answer to your fears in the workflow itself.

The Innovate on Purpose Workflow is built to streamline and focus your efforts from opportunity identification to commercialization. It does this by tying all market research to the goals of the organization and establishing priorities to fill the most important knowledge gaps first, and to keep resource expenditure under control with tight cycles of questions and experimentation. Innovation on purpose is *market research* on purpose. It's *analysis* on purpose. It's product development and commercialization and pricing and delivery on purpose.

My intent is to help you get successful solutions to market as quickly as possible, not keep you so busy doing market research you never get on with it. Following are four perspectives to keep in mind as we move into the last discovery-oriented phase of the workflow:

MARKET RESEARCH IS THE PROCESS OF INNOVATION, NOT WORK FOR ITS OWN SAKE.

As mentioned before, many organizations listen to their markets in sporadic, haphazard ways that disrupt what's seen as "our main job." But it's this very relationship with the market that's the reason to have a Market Strategist in an organization! Consider the process of closing the Knowledge Gaps in your Innovation Map an ongoing process that you devote a specific amount of time, money, and resources to every day. It *is* the innovation workflow.

GAINING MARKET KNOWLEDGE INCLUDES PRODUCT ITERATION.

Many of your hypotheses will be validated (or not) by taking wireframes, samples, options, A/B testing to the market. I visualize the gathering and use of market knowledge as a tight, upward spiral of questions and answers that progress toward a better and better product.

AVOID ANALYSIS PARALYSIS BY DEFINING "ENOUGH KNOWLEDGE" WITH A CONFIDENCE LANGUAGE.
Every organization has a unique risk tolerance and comfort level with the unknown. You as the Market Strategist should bring your cross-functional team together around a shared understanding of how much risk we're willing to take and then act on the knowledge when that threshold is achieved.

IF YOU DON'T KNOW WHY YOU'RE DOING THE RESEARCH, DON'T DO IT.
Too much market research is done by rote, without a specific goal in mind. Even when you're trying to learn something wholly new about your market, you should still have goals for the object and attribute you're investigating. The biggest source of analysis paralysis is research without a goal.

In short: don't paralyze yourselves with knowledge—*enable* yourselves with it. Now, we'll turn our attention to interpreting the knowledge you've gathered.

ACTION PLAN FOR CHAPTER 11

- Review the products and/or markets you're responsible for. Overall, what phase of the continuous conversation with your market are you in now? How are you focusing your innovation workflow to achieve it?

- Meet with your data scientists and market research team. What data streams and ongoing research projects could be honed to deliver richer insights? What could be stopped or replaced with something more effective?

- Meet with your product development team. How are you assessing feature priorities? How are you testing the build? Could these processes be streamlined or improved?

- Check the questions for your next survey against your goals for doing it. Will the tool as written get you the insights you're looking for?

Chapter 12
INTERPRET

Find Opportunities

· Investigate
· Experiment & Validate
· Interpret

ROADBLOCKS TO INTERPRETING
MARKET KNOWLEDGE

Back in Chapter 10, I said that market research was only the
beginning of your Innovation journey. Most organizations are
gathering a lot of market data, but they're not getting all the value
out of it that they should be. In my experience, there are several self–
inflicted roadblocks to companies getting full value out of the market
knowledge they gather. At this stage in your reading, they'll probably
look familiar:

LACK OF AGGREGATION

First, market knowledge isn't being aggregated to enable strategic
interpretation. There's no workflow; there's no acknowledgement
that it should be done. Each bit of market research completed is
treated as an event. Most organizations are ahead of the average if
they're aggregating like data points, such as NPS scores or cancellation
renewal codes. Nobody's bringing all of these insights together.

LACK OF OWNERSHIP

Second, there's ***no one in charge of*** creating an innovation strategy
or of folding interpreted market knowledge into the build, market,
price, and sell workflows. Product teams are focused on driving the
roadmap and grooming the backlog. Marketing looks at email open
rates and ad performance. Sales keeps an eye on contract renewals. But

nobody brings gaps and changes in the market together to examine them for opportunity. This is why the role of Market Strategist is so critical to innovation—you need someone who considers all facets of market need and all tools that are available to fill that need, within the context of an organizational strategy.

LACK OF PERSPECTIVE

Finally, there's the myth that groundbreaking new products are the brainchild of a genius, whole and entire. That a lightbulb goes on and those super-smart folks deliver the Next Big Thing. As a result, product teams think it's *okay* that the first two roadblocks exist. They *feel shy of data interpretation because it seems futile* and, frankly, somewhat pretentious. But Disruptive Innovation is more often realized through methodical connecting of indicators in the market, testing in rapid cycles, and learning from the effort.

There is art and perhaps even a bit of magic to innovation, but inspiration and opportunities can be found by looking for them. Whether you are trying to teach an old product new tricks or redefine an entire market, following are a few lenses through which to synthesize and interpret the market knowledge you've gathered in order to spark your innovation efforts.

INTERPRETING MARKET KNOWLEDGE BY INNOVATION PHASE

A good way to decide what techniques you should use when interpreting market knowledge is to ask yourself, "What phase of the Innovation Spectrum am I trying to achieve?"

There's very little interpretation needed in the Reactive Innovation phase. Because the objective here is to build what users are asking for, your focus will be on understanding their priorities and matching your product development efforts to them. Responsive Innovation is only slightly more complex, in that you're investigating additional objects such as buyers, competitors, influencers, and governors. You'll need to aggregate this information and deliver a balance of benefits for each.

OBJECTS OF OPPORTUNITY RESEARCH ALIGN WITH PHASE OF INNOVATION DESIRED

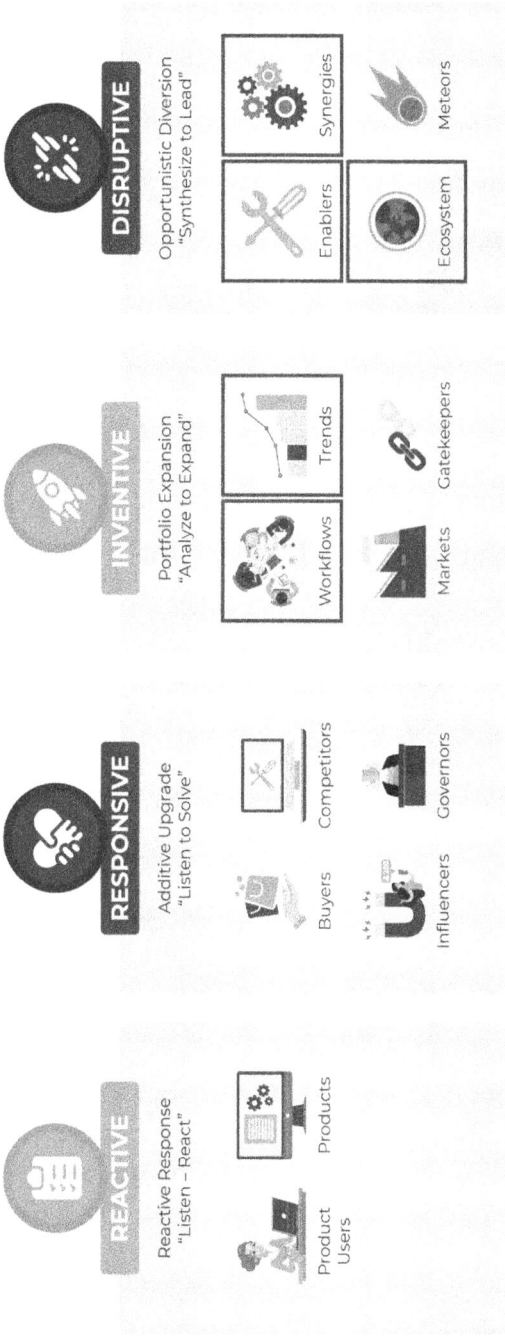

REACTIVE

Reactive Response
"Listen – React"

Product Users

Products

"PRIORITIZE"

RESPONSIVE

Additive Upgrade
"Listen to Solve"

Buyers

Competitors

Influencers

Governors

"AGGREGATE"

INVENTIVE

Portfolio Expansion
"Analyze to Expand"

Workflows

Trends

Markets

Gatekeepers

"COMPARE"

DISRUPTIVE

Opportunistic Diversion
"Synthesize to Lead"

Synergies

Meteors

Enablers

Ecosystem

"EXTRAPOLATE"

Objects of Opportunity Research Align with Phase of Innovation Desired

Where interpretation becomes very important is if you aspire to Inventive Innovation and beyond. For Disruptive Innovation, data synthesis, interpretation, hypothesizing, and experimentation are key differentiators and must-haves. What are some methods you can use to turn market knowledge into innovation?

INVENTIVE INNOVATION: CHANGES IN WORKFLOWS AND TRENDS

Inventive Innovation is the phase where organizations aspire to lead their existing markets and grow through expansion into new markets. One requirement is that you begin gathering data about objects *over time*.

Two good examples of how to go looking for this phase of innovation include understanding what's changing in the workflow you serve and changes to each persona within it.

WHY DO WORKFLOWS CHANGE?

Workflows are the processes a group of individuals use to produce a result. The result could be, for example, the purchase of a new accounting system or the use of that accounting system. In a home, it could be how a family chooses its media subscriptions or how they choose what to watch on a Wednesday evening. You need to understand the workflows you're part of, and the data you collect will help with that. It should also enable you to see the ways those workflows are changing and how you should change with them and perhaps even lead the way to a new and better way of getting something done. Here are a few catalysts to workflow changes:

- *Workflows can change due to logistics.* In the early 2020s, many workforces (and therefore workflows) were highly distributed because of COVID lockdowns, working individually from their own homes yet still needing to collaborate. Online collaboration required individual contributors to turn their workspaces into small studios, complete with microphones and special lighting. There were opportunities for a wide variety of suppliers to expand their product lines and markets—so much needed to be automated at a micro-office level. As of this writing, organizations are moving back toward more co-location, but not to pre-COVID levels. How could Zoom and other providers change the way they enable the logistics of these evolving teams?

- *Workflows can change due to technology.* Virtual/augmented reality, robotic process automation, and other technologies are changing our expectations of how customer service, manufacturing, and even health care are delivered. They're changing our very definition of the "real world." There is plenty of opportunity to compete by building these enabling technologies, but it's also worth considering how these enabling technologies will change the way your markets expect to buy, sell, and otherwise interact with you. If not, your product engineer may find themselves talking to a robot.

- *Workflows change due to cultural expectations.* More than one of my clients has mentioned running into sustainability officers in buying workflows. These folks are asking questions about LEED, DGNB, or Green Star ratings, recycling efforts, and carbon footprints. Are your sales teams ready for that part of the conversation? Does your marketing address those specific expectations? The buying workflow in your markets may have changed to include this new facet of decision making.

How buying decisions are made, how products are built, and how they're used by people—or not—are all changing right now and will continue to do so over time. The velocity for each of these aspects could be very slow or whiplash-fast. Interpreting the data that you gather to identify those changes should enable you to meet and even lead them. That brings us to the next lens you should look through to find innovation opportunities: trend analysis.

HOW TRENDS IMPACT INNOVATION

Beyond year-over-year growth of revenue or renewals, very few organizations perform trend analysis on their markets. Yet market shifts are an excellent opportunity to stay ahead of the competition and stay relevant in the eyes of your customers. Every market changes over time, which means how you serve that market must change too.

EVOLUTION OF DEMOGRAPHIC CHARACTERISTICS

I'm going to use the example of actual versus perceived characteristics of age to illustrate the value of keeping on top of trends. Not because age is the only demographic trend that can bite you but because it's one that bites a lot of product teams, especially those in tech.

The people building, selling, and marketing most of the stuff I use could be my kids. They do market research and build personas to help them understand me and people older than me, but sometimes the results are terrifying.

This is because "old" has changed. My 80-year-old father does his banking online. My 93-year-old aunt regularly Skyped with her great-grandchildren—a decade ago. They didn't grow up with smartphones and tablets and the Internet of things, but they've been using them for a long time. In a lot of cases, they *invented* them.

On the other side of the spectrum, 11-year-olds today are very different than 11-year-olds of 2015, even 2020. If you've been building video games since 2014, you can't fall back on your own memories of what was fun, how you wanted to play, or what your parents would pay for when you were that age, even if you haven't hit 30.

Some demographic-specific characteristics pervade, such as increasing mobility challenges and retirement from work, for those over 65, but others most definitely do not. Keeping track of what it means to be "an age" is important if your product serves that age group; this includes everything from daycare to Medicare. What's changing about a demographic and what's not is something you need to know to build not just responsive products but also marketing, positioning, pricing, and distribution methods. I call this the phenomenon of the **maintaining persona**, or serving a market that's defined by age.

On the flip side, you may be faced with innovating for an **aging persona**, in other words a demographic set that stays with you as they age. Everything from clothing brands to social media sites can start out with an audience of 20-something loyalists that stay with them through the decades, but don't catch on with the latest batch of 20-somethings. This is the aging persona.

The graphic below is a good example of these two demographic trends and some of the questions you may come up with during data interpretation. Chances are, your interpretation will need some validation before full-scale innovation!

CHANGING DEMOGRAPHICS – WHAT DO THEY MEAN FOR INNOVATION?

Changing Demographics— What Do They Mean for Innovation?

HYPOTHESES & BRAINSTORMING

AGING PERSONA
Average age of the persona buying/ using your product is increasing.

Will messaging emphasizing "experience," "transition," "legacy" resonate with this persona?

If B2B, investigate whether this role is disappearing and what, if anything, is taking its place.

Should we modify our offering to maintain loyalty? E.g., larger font size to allow for vision challenges.

Can we modify our offering to increase the customer base? E.g., new branding, updated integrations.

MAINTAINING PERSONA
Average age of the persona buying/ using your product stays the same

What new themes should we emphasize so our marketing resonates with this persona?

Is the way this persona wants to buy our product or communicate with customer service changing?

Could onboarding be a key differentiator for us?

Should we modify our offering to maintain loyalty? E.g., integrate a smartphone app.

Can we modify our offering to increase the customer base? E.g., larger font size to allow for vision challenges

Finally, I'll suggest three additional techniques to interpret the market data you've gathered to achieve Disruptive Innovation.

DISRUPTIVE INNOVATION: CONVERGENT AND DIVERGENT THINKING, ENABLERS, AND 360-DEGREE IDEATION

Have you ever used a coin as a screwdriver? An untwisted clothes hanger as a hook? These may not seem like Disruptive Innovation, but the same principles apply to these workarounds as to pharmaceutical companies marketing prescription drugs directly to potential patients and the invention of the iPod.

To achieve Disruptive Innovation, you'll go beyond your market, beyond the product, and likely beyond your comfort zone. The Disruptive Innovation phase isn't for every organization, but if you've chosen to do go for it, here are some ways of thinking that can help.

DISRUPTIVE INNOVATION REQUIRES AN EXPANDED UNDERSTANDING

You've heard it before: the parable of the blind men and the elephant. Originating in India and tweaked through time and custom, it's a story of a group of blindfolded people that encounter an elephant and attempt to understand what it is by touching it. Because each person can only "see" a small piece of the elephant, each one guesses wrong about what they're touching—and each one guesses something *entirely different* based on the piece of the elephant they're touching. At last, a sighted person enters the group and describes the entire elephant by unifying the various perspectives. While the perspective of each of the blindfolded people was true, it wasn't the *entire* truth.

In our context, we know that a single view of a buyer, user, or market can be true but that trying to build innovation on that single facet of knowledge is inherently limited by its failure to account for additional facets, the evolution of each facet, or a totality of facets. In other words, I want you looking at the whole elephant.

To find inspiration for Disruptive Innovation, look not only at your market but also the world around it. Is there new enabling technology you could bring to bear on an old problem? Have shifting regulations opened an opportunity previously blocked? Are people today buying and using consumer products in a way you could copy?

To methodically try to innovate at this level, you need to explore the market in a much different way than most organizations do. First, you need to expand what you think of as objects, or targets of study, beyond personas, markets, and competitors and even beyond regulatory bodies, influencers, and trends. You'll cast a wide net for this type of inspiration and bring what you learn together in a different type of brainstorming than any other innovation phase. To visualize this, we'll use the classic concepts of convergent and divergent thought.

CONVERGENT AND DIVERGENT THOUGHT IN DISRUPTIVE INNOVATION

The terms "convergent" and "divergent thinking" were coined in 1956 by American psychologist J.P. Guilford to describe different ways people figure out how to solve problems. We're going to use Guilford's work to examine how Disruptive Innovation is different from less extreme phases on the Spectrum.

Convergent thinking is when someone moves from having a problem to solving it with the most direct, well-defined option. It's quick, delivers a solution that's unambiguous, and involves linear thinking. A good example might be if the dishwasher breaks, you call the serviceperson to fix it. This type of thinking is best suited for tasks, including solution creation, that involve logic as opposed to creativity. Reactive and Responsive Innovation phases can be achieved largely through convergent thinking because they both involve improving existing solutions with existing tools.

Divergent thinking is non-linear and involves brainstorming solutions that may take advantage of unrelated tools, technology, and societal norms. With this type of thinking, you can generate ideas and develop multiple possible solutions to a problem. That said, the goal is the same as convergent thinking—to arrive at the best solution. If you used divergent thinking on the dishwasher example, you may also consider doing dishes by hand, hiring someone to do so, or buying disposable plates. For product teams looking for Inventive or Disruptive Innovation, it's the reason you expand your market knowledge beyond buyers, users, and competitors to awareness of technology advances outside your industry and consumer expectations that might bleed over into their workday assumptions, for example.

> **Convergent and divergent thinking are opposite methods of thinking used to come up with the best solution to a problem.**
>
> **For a Market Strategist, convergent thinking is best used to deliver Reactive or Responsive Innovation because its goal is to find the most obvious and logical solution to a market problem based on current knowledge.**
>
> **For those aspiring to Inventive or Disruptive Innovation, divergent thinking enables better results because you take into account indirect options to solve the problem.**

Much of market-driven innovation is built on solid convergent thinking, but divergent thinking, often called creative problem-solving, is key to uncovering Disruptive Innovation opportunities.

For a Market Strategist, **convergent thinking** is the workflow of bringing together facts and data that point to a market need you can define and a solution you can create. You would typically use convergent thinking in Reactive and Responsive Innovation. A basic workflow for convergent thinking in product innovation is shown on the left side of the graphic below:

On the right side of the graphic is an idea of how **divergent thinking** works. Instead of bringing together increasing evidence that a particular solution is needed by the market, divergent thinking asks questions about what could be, or questions the need to accept a current situation that's subpar. The two-step process starts with a what-if or a "so what?" question and explores all the tools that might be used to resolve it.

INNOVATION – CONVERGENT THINKING

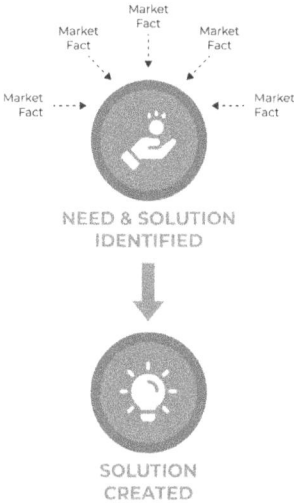

Market Fact · Market Fact · Market Fact · Market Fact · Market Fact · Market Fact

NEED & SOLUTION IDENTIFIED

SOLUTION CREATED

INNOVATION – DIVERGENT THINKING

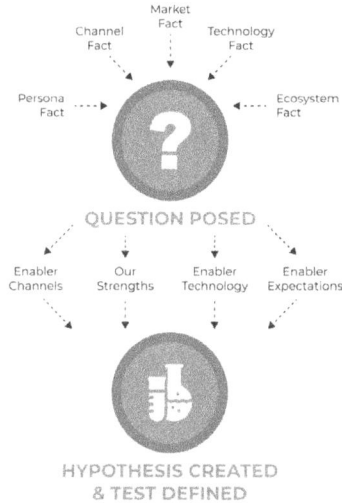

Channel Fact · Market Fact · Technology Fact · Persona Fact · Ecosystem Fact

QUESTION POSED

Enabler Channels · Our Strengths · Enabler Technology · Enabler Expectations

HYPOTHESIS CREATED & TEST DEFINED

Innovation—Convergent & Divergent Thinking

What if marketing tools could be created in multiple languages simultaneously? What if a sales rep didn't need to speak Spanish to be understood by a Spanish-speaking client? Brainstorming the many possible solutions to the question includes a wide-ranging toolkit of technology advances and regulatory changes, observations of how market players behave in unrelated workflows, and how all of this could be brought together to create a wholly new solution—perhaps after having articulated a previously undefined need.

If divergent thinking brings into account all options to solve a problem, its opposite number is 360-Degree Ideation, which encourages the cross-functional team to examine all possible "component" options to solve a market problem.

360-DEGREE IDEATION POINTS OUT ALL INNOVATION POSSIBILITIES

Too often, the ideation phase of innovation is attended by a really exclusive crowd. Too often, this leads teams with Disruptive Innovation goals to build disappointingly incremental, reactive solutions. Throughout this book, I've added elements to your workflow to change that. We'll ideate using **360-Degree Ideation** now to leverage that work.

Market Strategists hear all sorts of frustrations and unfilled desires when they listen to the market. Here are a few I've personally heard in my career:

> *"It would be great if this worked like my music app does!"*

> *"I wish you'd let me just get the dang white paper without giving you all my contact information—I'm not ready to talk to a sales rep yet!"*

> *"Why do I have to commit to a one-year plan? I could buy today if you'd let me go month-to-month."*

> *"This 27-page contract is a little, uh—much. I'll never get it through my legal department!"*

Clearly, there are some frustrations here that could be solved by any market-driven organization. ***But they don't even make it to the table.***

Why? Because the ideation sessions in most tech companies include three groups: IT, UX, and product. When was the last time someone from your contracts department was invited? Marcomm? Pricing? What was the last non-product innovation you implemented to solve a market problem.

360-Degree Ideation brings together subject-matter experts from different parts of your organization to brainstorm solution options based on the broad understanding of the market you've achieved by investigating a wide variety of objects intentionally to enable Disruptive Innovation.

In order to deliver Disruptive Innovation you have to consider all types of innovation. Your market considers everything you do for them "product." It's time we did too.

So interpretation to disrupt a market should combine divergent thinking with a 360-Degree Ideation and discussion of your market knowledge. This discussion goes beyond "a product" to break down barriers of delivery, pricing, marketing, and selling.

DIVERGENT THINKING AND 360-DEGREE IDEATION TAKE THE MYSTERY OUT OF "MIRACLE" PRODUCTS

Early in the book, I mentioned that some product teams give up on innovation entirely because they believe really cool "miracle" products resulted from the personal brainwaves of a few big-time geniuses. But if you look at the market signals that were available between old and new—it's apparent that these "miracle" products were really a product of some divergent thinking and 360-Degree Ideation:

"MIRACLE INNOVATION" CAN BE ACHIEVED THROUGH SYNTHESIS AND EXTRAPOLATION

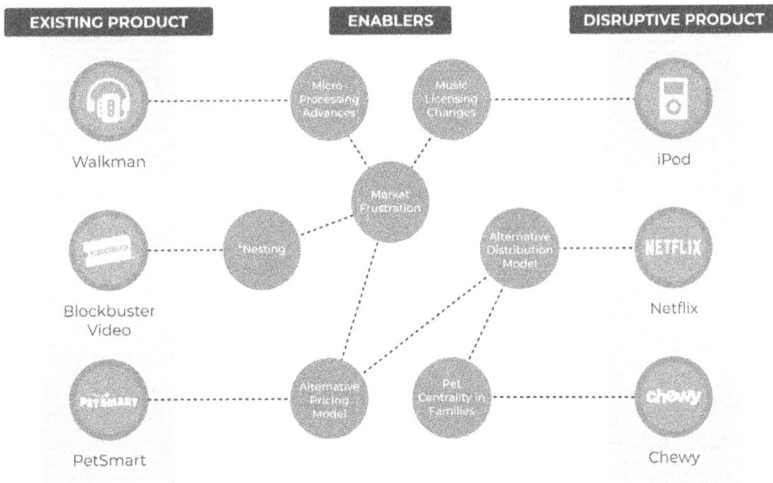

EXISTING PRODUCT	ENABLERS	DISRUPTIVE PRODUCT
Walkman	Micro Processing Advances / Music Licensing Changes	iPod
Blockbuster Video	"Nesting" / Market Frustration / Alternative Distribution Model	Netflix
PetSmart	Alternative Pricing Model / Pet Centrality in Families	Chewy

"Miracle Innovation" Can be Achieved Through Synthesis and Extrapolation

For example, the iPod wasn't a product breakthrough, it was a *technology* breakthrough that enabled a reboot (one of many in my lifetime) of the way we listened to music and a *licensing* breakthrough that enabled the sale of individual songs to stream.

Netflix had the same product Blockbuster did when they achieved market dominance—movies and TV shows on hard media like DVDs that could be rented and returned. But they took advantage of the fact that people hated Blockbuster's late fees and were increasingly relying on subscriptions for their at-home media. The pricing model, as much as all the other choices they made, enabled Netflix revolution of the video market.

PetSmart—a whole store devoted to the needs of pets, was only possible because of the increased pet centrality (and disposable income) in family life, but that centrality (and money) plus subscription models in the consumer world led to Chewy and Barkbox.

I'm not saying the thought process illustrated above was exactly how Ryan Cohen and Michael Day were inspired to build out the pet-subscription box Chewy. What I *am* saying is that, if you want to go looking for opportunities for Disruptive Innovation instead of waiting to stumble onto one, this is a good way to do it.

Use the data you've gathered on that wide variety of objects, and brainstorm using divergent thinking and 360-Degree Ideation, and you'll create some pretty amazing hypotheses!

Using all the insights from our Knowledge Center, a cross-functional team can brainstorm solution ideas from many perspectives and come up with something to test.

ACTION PLAN FOR CHAPTER 12

- List your roadblocks to interpreting market knowledge. How could you eliminate them?

- What innovation phase are you aspiring to? What techniques and knowledge should you add to your interpretation workflow in order to achieve it?

- Identify one market problem and experiment with the two ways of thinking—convergent and divergent—to solve it. Which makes more sense in your organization?

- Make a list of cross-functional team members to invite to your next ideation session.

Chapter 13
COMMUNICATE TO EMPOWER

Communicate to Empower

Congratulations! You've worked the innovation process and discovered some exciting opportunities in your market, possibly without having to do much additional market research. But to bring these opportunities to market, you're not the only one who needs this knowledge. Too often, market research is performed, and products built, seemingly without the first informing the second. It's time to establish a last step in mining your existing awesomeness, the step of communicating what *you* know to inspire the rest of the team.

In this section, I'll outline your communication responsibilities as the Market Strategist and offer some best practices to create a communication plan that's both effective and realistic given your existing time and resources. Along the way, I'll remind you of several communication tools you've already created.

WHAT WILL YOU COMMUNICATE?
Market Strategists have always been the link between the market and the business, but often our communications—like our market research—are overly focused on what we're building for users right now, instead of what we're earning about the market for the future.

There's nothing wrong with conveying traditional roadmap and release plan information, but it limits not only your ability to build a forward-thinking product, it also doesn't offer insight into opportunities for the future, or beyond the product itself. Just as we've spent the last several chapters incorporating what you know about more than just users and your products, I'll ask you to communicate more to your teams as well.

ESTABLISH SCOPE AND CORPORATE FIT WITH YOUR CONTRIBUTING VISION AND STRATEGY

A good first step is to make sure your stakeholder network understands what knowledge you're responsible for gathering. In larger organizations, shared services such as IT or public relations likely work with several Market Strategists, each representing a different set of products, markets, priorities, and goals that contribute to achieving the overall company strategy. You need to focus their thinking on the strategy for your scope of responsibility so they can respond to those unique objectives. In Chapter 4, "Aligning with Leadership," you worked with leadership to create the vision and strategy specifically for your scope of responsibility: the contributing vision and strategy. Use this visual and the detail behind it to set the stage for your cross-functional team.

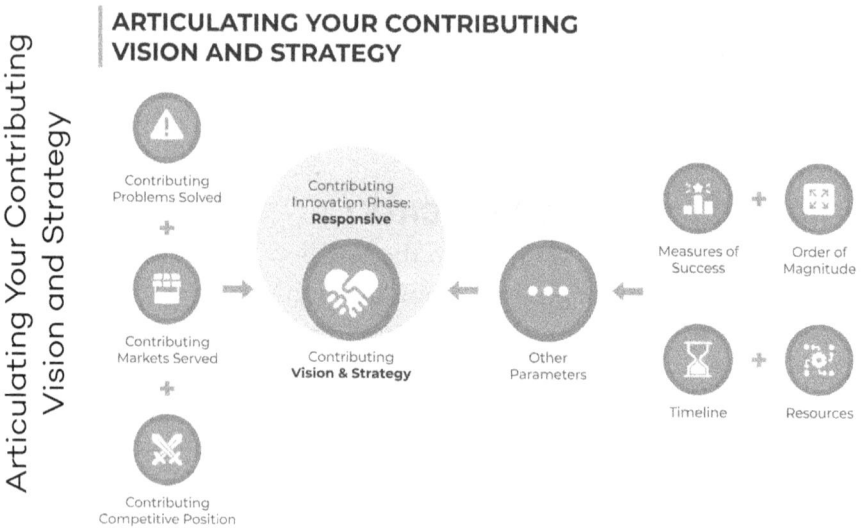

Articulating Your Contributing Vision and Strategy

ARTICULATING YOUR CONTRIBUTING VISION AND STRATEGY

Contributing Problems Solved
+
Contributing Markets Served
+
Contributing Competitive Position

Contributing Innovation Phase: **Responsive**
Contributing **Vision & Strategy**

Other Parameters

Measures of Success
+
Order of Magnitude

Timeline
+
Resources

Most organizations aren't used to looking at the market beyond buyer and user personas, so you'll have to set the stage for a larger conversation and awareness with your team. I recommend using objects, attributes, and sources to define the scope of innovation, and an excellent visual to do this is the Market Strategist's Innovation Map. This tool was introduced in Chapter 6, and a LegalcoEG example shown. Below is another example of a more-detailed Innovation Map. As you can see in this example, it quickly gives you a look at the objects, attributes, and sources the MyPatentSoftware team is responsible for:

COMMUNICATE CONTEXT BY BUILDING PERSONAS—AND BEYOND

Too often, teams dive into creating positioning, solution ideation meetings, user stories, and other "response" documents when they don't have a high-level understanding of the world they serve.

Several years ago, I was advising a team on launch strategy for a new product. The team was all for launching into their biggest market, which was publishers of print magazines. The buyers and users in that market were well known, and the sales team had existing contacts—it seemed logical if you looked only as far as buyers and users.

Unfortunately, that market was dying. The opportunity for my client was dying with it. If you looked at market health, it was obviously a bad bet to chase any new business in the print publishing world at that time. The client knew the buyers and users in the market, and they also knew that the answer to any increase in spend was going to be no. They just weren't connecting the dots.

You already know it's important for teams to have context beyond buyers and users. The question at hand is, How do you *communicate* that context so your teams can make better decisions?

A persona-like format is great for communicating the context of more than just buyers and users; consider building "personas," or contextual summaries, on whatever objects you chose to investigate in our earlier chapters.

EXAMPLE INNOVATION MAP: MyPatientSoftware, Inc./
IP Law Firms/Managing Partner

EXAMPLE INNOVATION MAP: MYPATENTSOFTWARE, INC/IP LAW FIRMS/MANAGING PARTNER

Listening to a market requires attention on many levels. A good first step toward building a listening strategy is to identify the constellation of voices in each market by building a Innovation Map. Below is an example using our mythical LegalcoEC's MyPatentSoftware product.

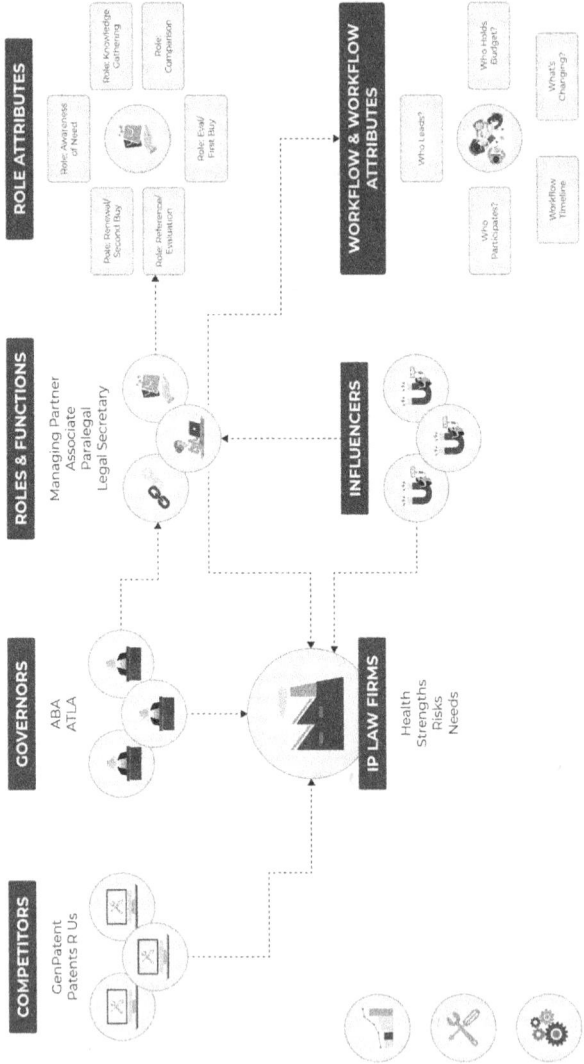

COMPETITORS
GenPatent
Patents R Us

GOVERNORS
ABA
ATLA

ROLES & FUNCTIONS
Managing Partner
Associate
Paralegal
Legal Secretary

ROLE ATTRIBUTES
Role: Awareness of Need
Role: Knowledge Gathering
Role: Competition
Role: Eval/First Buy
Role: Removal/Second Buy
Role: Preferential Evaluation

INFLUENCERS

IP LAW FIRMS
Health
Strengths
Risks
Needs

WORKFLOW & WORKFLOW ATTRIBUTES
Who Leads?
Who Holds Budget?
Who Participates?
Workflow Timeline
What's Changing?

Back in Chapter 6, I introduced the concept of objects. While they include personas like Vivienne, managing partner of a law firm in Canada, they could also include product categories ("Case management software"), markets you serve and compete in ("Law firms in the US"), as well as influencers, governors, and gatekeepers to name a few. Creating a persona for Davina, who uses your kind of product, is a logical step toward creating context with your team. But what about building a shared understanding of the market and the agencies that govern it, the influencers who will help them choose your solution, or the emerging trends and changes in all of the above?

There are good persona templates available from a wide variety of sources, so I'm not including one here. Rather, below are some suggestions for what to include for non-traditional objects of persona building:

CREATE OBJECT "PERSONAS"

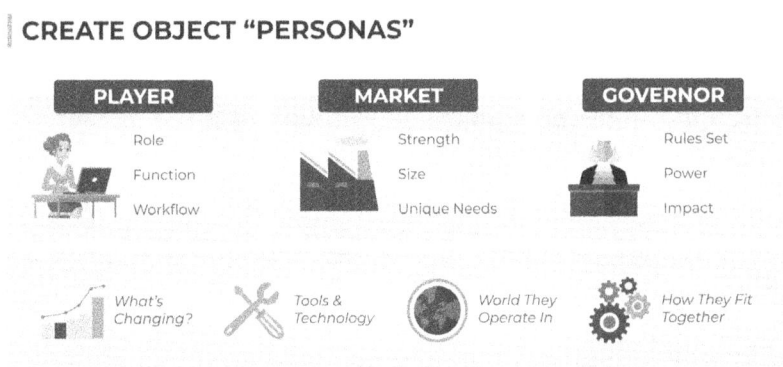

PLAYER	MARKET	GOVERNOR
Role	Strength	Rules Set
Function	Size	Power
Workflow	Unique Needs	Impact
What's Changing?	Tools & Technology	World They Operate In	How They Fit Together

As mentioned earlier, the need for context on one or many objects is defined by your appetite for disruption. Reactive Innovation requires much less market context than Disruptive Innovation.

As you can see, we're drilling down from high-level insights to more actionable information. The last step in your responsibility for insight is to provide some concrete examples of market opportunity.

ENABLE ACTION WITH OPPORTUNITY STATEMENTS

The most tactical responsibility of the Market Strategist is to share opportunities you've identified—for experimentation to existing product sunset—in a way that's actionable by the team. In many ways, this is telling the story of the opportunity you have in the market, so let's call them opportunity statements.

Opportunity statements are statements of needs, desires, influences, or gaps felt by your organization and the objects that you act on to succeed in the market. Generally, they illuminate the attributes of an object, such as "Growth trends for the law firm market" or "Professional constraints experienced by Javier, law firm manager in Mexico." Data you already gather through sources such as surveys or machine learning and aggregated into Knowledge Statements help you form opportunity statements, and the strength of that statement will guide your decision to innovate.

> **Opportunity statements express an opportunity your company could go after. Often inspired by Knowledge Statements, opportunity statements may be actionable as market experiments or even product or other direct market action.**

Ideal knowledge statements include persona/market need, impact of need on the persona, impact of need on us, desired outcome, and any constraints such as compliance or deadlines. For example:

Javier, Mexico law firm manager, feels unimportant to us because our customer service chatbot and telephone reps don't speak Spanish. This gap also causes Javier wasted time in problem resolution—if his problem is resolved at all. If we had Spanish-language customer service resources, our NPS would improve. We've calculated a one percentage-point increase in NPS to equal $1.5M in top-line renewal revenue.

Here's an example for a less traditional "persona": a governor of the legal market in Mexico, the General Professions Bureau:

Licenses issued to lawyers by the General Professions Bureau are currently valid in all states in Mexico; licensed lawyers can practice anywhere. However, the Federal government has approved the Bureau's proposal to offer specialized legal licenses by both area of practice and state, to be implemented in September 2023.[1]

This creates an opportunity for us to offer new/expanded advertising products in our law firm marketing product suite. We should also consider updates to our pricing and our own marketing messages to the law firms to reflect their need for multiple or bundled offerings.

Also note that this opportunity statement goes beyond the product to include opportunities to change pricing and marketing messages.

YOUR ROLE IN COMMUNICATING MARKET KNOWLEDGE

If you're the Market Strategist for your product, you are the person responsible for making sure everyone in your organization knows enough about the market to do their jobs well. But what does that mean?

TAKE IT SERIOUSLY

I once had a product team colleague actually say these words to me, "I'm too busy gathering all this market data to be worried about communicating it!"

Without effective communication, gathering market data is a waste of time. The point isn't to *do* surveys and focus groups or pay data scientists a lot of money to create analyses. The point is for everyone in the organization to come together around two guideposts: what

---

[1] Note that is an example for illustrative purposes only; there is no such proposal or plan to my knowledge.

your market needs and who you want to be. Only this knowledge enables them to *respond* with incredible products at the right price, marketed effectively, and sold the way the market wants to buy. To do that, the Market Strategist must communicate effectively about the market and opportunities—*before* the leap to building user stories, customer service apps, or value statements.

So take this part of your responsibility seriously by allocating time for it. Yes, it will take away from time learning about the market. But unused knowledge is a wasted effort. How much time should you allocate to communication? That depends on your organizational complexity, market velocity, and a wide variety of variables—but not on your total amount of work time. A random-but-effective guesstimate I used was 75% of time spent investigating, 25% communicating.

RECONCILE GAPS AND INCONSISTENCIES
ACROSS CONTRIBUTING VISIONS & STRATEGIES

Have you ever run into a situation where your product appears to compete with another—from inside your own company? Or had a marketer push your product launch to the back burner to work on something that was "a higher priority?" Certainly, we've heard the lament that there's more to do than there are resources to do it—and probably said it ourselves! These stem from a lack of reconciliation across product and shared services teams and cause all sorts of mischief.

The reasons why this reconciliation *doesn't* happen could fill another book, so I won't dwell on them here. However, I strongly recommend that you get together with any other Market Strategists that use the same shared services teams you do, and do three things:

- *Identify the gaps and inconsistencies:* Does it seem that there's no one responsible for some area of the corporate vision and strategy? Or maybe there are two products being worked on that compete with each other, and no one knows why. Get together with your team and get a list of questions together about anything weird.

- *Resolve these through discussion and leadership input:* Get clarity from leadership about any gaps or inconsistencies that don't make sense to you. That said, understand and accept that some weirdness across product lines and company divisions will always exist.

- *Plan to meet regularly with the other Market Strategists:* Be sure to include representatives of shared services teams. Work together to reconcile priorities, make trade-offs, and escalate issues based on the resource constraints of the shared teams. Meet monthly to start and then quarterly as alignment solidifies.

Taking these steps won't eliminate all confusion and inconsistencies, but it will reduce them. And you'll be amazed how just acknowledging that reconciliation isn't perfect, and that we'll have to work through that, will improve a team's outlook and encourage them to work together to resolve what they can.

MANAGE EXPECTATIONS

The goal with managing expectations is to let your team know how much real data backs up any assertion you're making about the objects and attributes you've got your eye on.

We already discussed how to do this by creating a Confidence Language with the confidence levels "Believed, Perceived, and Recognized," or "Red/Yellow/Green." Leverage the Confidence Language you built earlier on an ongoing basis to quickly give teams a sense for how much risk they're taking when they respond to the knowledge you give them, and get them used to using the same language in their own assertions.

CREATE A CONFIDENCE LANGUAGE

Telegraph level of certainty to the team.

Establish common understanding.

- Do we have enough knowledge to act?
- If not, what else do we need to feel comfortable?
- Create a list of hypotheses to validate.

Establish a language for market change.

UNKNOWN (GRAY)
"We should know this, but don't."

BELIEVED (RED)
"We believe this."

PERCEIVED (YELLOW)
"We have some data on this."

RECOGNIZED (GREEN)
"This meets our data threshold."

TRANSITIONING (BLUE)
"This is changing."

I'll use the examples of Javier and the General Professions Board of Mexico to illustrate:

> *Javier, Mexico law firm manager, feels unimportant to us because our customer service chatbot and telephone reps don't speak Spanish. This gap also causes Javier wasted time in problem resolution—if his problem is resolved at all. If we had Spanish-language customer service resources, our NPS would improve. We've calculated a one percentage-point increase in NPS to equal $1.5M in top-line renewal revenue.*

This opportunity is Perceived—we have a couple of complaints lodged via chatbot, and the topic came up in a focus group last month.

> *Licenses issued by the General Professions Bureau are currently valid in all states in Mexico; licensed lawyers can practice anywhere. However, the Federal government has approved the Bureau's proposal to offer specialized legal licenses by both area of practice and state, to be implemented in September 2023.*

This creates an opportunity for us to offer new/expanded advertising products in our law firm marketing product suite. We should also consider updates to our pricing and our own marketing messages to the law firms to reflect their need for multiple or bundled offerings.

This opportunity is Recognized—it was reported by the Federal government of Mexico and the Bureau in a joint press release last week.

CITE YOUR SOURCES

The Confidence Language is a good, quick indicator of the strength of any knowledge you're asking teams to respond to, but sometimes more context is necessary. If a team member wants the backstory on your rating of "Perceived," be ready to cite the data sources. Particularly in early days of building your relationship with a new team, you're also building trust. Your team has a right to get comfortable with the actual data, so the more transparent you are, the better. As you can see in the above examples, I anticipated the need and included the data highlights in my communications.

With these responsibilities and deliverables in mind, it's time to create a communication plan.

SOME THOUGHTS ON BUILDING A COMMUNICATION PLAN

As with persona templates; communication tools and best practices are available, so I won't attempt to make up something better here. With that said, there are a few often overlooked techniques that I've learned from experience can improve *any* communication plan, and I offer those here.

HAVE A PLAN

If it seems like I've morphed into Captain Obvious—I agree. But *not a single team* I've advised or been hired to lead had a stated plan to communicate market knowledge to their cross-functional team in place when I arrived. And I'll admit it: I didn't build, request, or suggest such a plan until very late in my own career.

Yes, these teams had epics and user stories—seen by the developers working on a specific product. They had personas—that hadn't been updated in three years, gathering virtual dust in a forgotten folder in Teams. They also had status meetings, updates, and reports of all types: focused on yesterday's work, technical roadmap updates, and projected revenue figures.

There's nothing wrong with any of those tools or types of communication. However, without a plan to deliver the particular insights we've uncovered throughout this book, you waste the knowledge you have and doom your team to Reactive Innovation status forever. So, create a communication plan—any communication plan—and stick to it. Don't wait for more resources or money; base your plan on today's resources. Not much? Fair enough; but if you've got a plan, you'll use whatever you have more wisely than if you don't.

Your plan should include messaging goals, tools, and time allocations for the following:

- *Onboarding for new employees* (**all** *new employees*) with a baseline knowledge of your innovation goals, objects, or targets of the innovation and an overview of what's known about those objects today.
- *Change notification.* When regulations, user goals, or market health changes, everyone needs to know.
- *Scope of innovation goals.* Use your contributing vision and strategy, Innovation Spectrum, and Market Map to illustrate your focus at the intersection of what your market needs and who you want to be.
- *Gaps and backlog.* You'll never know everything, and what's true today will change at some point. Make sure the organization knows you've got a plan to stay on top of things as well as fill any current knowledge gaps.

LEVERAGE EXISTING COMMUNICATION TOOLS, MEETINGS, AND MEDIA

Throughout this book, we've created visualizations and summaries that reflect your innovation process and the knowledge you're gleaning from that process. Many of those tools can be used not just to gain buy-in for future projects but to communicate insight to a team on an ongoing basis. In keeping with our goals to maximize impact from minimal effort and use what already exists, I recommend repurposing those tools wherever you can. I refer back to a few of them throughout this section as examples.

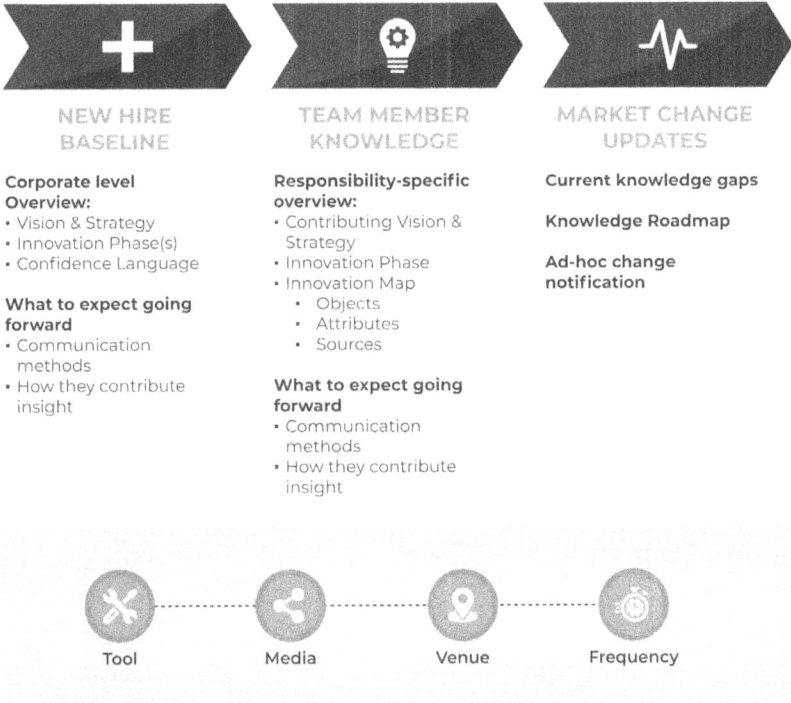

THE MARKET STRATEGIST DELIVERS INSIGHT TO THE TEAM

NEW HIRE BASELINE

Corporate level Overview:
• Vision & Strategy
• Innovation Phase(s)
• Confidence Language

What to expect going forward
• Communication methods
• How they contribute insight

TEAM MEMBER KNOWLEDGE

Responsibility-specific overview:
• Contributing Vision & Strategy
• Innovation Phase
• Innovation Map
   • Objects
   • Attributes
   • Sources

What to expect going forward
• Communication methods
• How they contribute insight

MARKET CHANGE UPDATES

Current knowledge gaps

Knowledge Roadmap

Ad-hoc change notification

Tool   Media   Venue   Frequency

I also encourage you to leverage the meetings, status reports, and communication tools leadership already expects you to deliver. If you have a persona template, Teams channel, or monthly all-hands meeting, embed the market knowledge you're delivering in them instead of building something entirely new.

DELIVER MESSAGING IN LAYERS AND REPEAT, REPEAT, REPEAT

When you use communication channels and tools the team is already interacting with, you repeat your message—and that's a *good* thing. For you, it means you can put together a sort of baseline messaging content and tweak it just a little for each source. Unless something changes, you will keep using the same messaging over and over. This not only helps you communicate more with fewer resources, it delivers the repetition that builds familiarity with the message.

BUILD FOR YOUR AUDIENCE

Where you *do* have to build net-new communications, do so with your audience in mind. Very few of us really want to read that detailed, 24-page PDF. Actually—nobody wants to read it. Ask yourself: What kind of media will my teams find engaging enough to sit through? When's the best time to get the message across? Fit delivery of new media through expanded workflows in as your time allows, prioritizing teams you may have left out of communications in the past but who could have a big impact on innovation such as pricing, data science/research teams, and legal. Here are some ideas:

- *Begin each meeting with a customer story.* Whether it's an existing customer, a competitor customer, or even someone who hasn't bought yet, remind everyone why you're here. A CEO I know has every team do this at every meeting. Meetings about hiring. Meetings about office supplies. Meetings about accounts receivable. Every meeting reminds every team member why they're there.

- *Create low-budget video and podcasts.* DIY media works with your budget and delivers the type of media most of us want to interact with.
- *Put the message where the eyes are.* Screen savers, your internal signature line in emails—wherever there's real estate to get the word out.
- *Ask marcomm.* They message to hard-to-reach prospects every day. I'll bet they have some great ideas you can steal for internal messaging.

ACTION PLAN FOR CHAPTER 13

- Compare strategic direction and tactical insights. How should you communicate those differently?
- Create an opportunity statement. How is it different from the Knowledge Statements you created in earlier chapters?
- Set aside a portion of each week for communication, and make a list of the existing tools, meetings, and other communication opportunities that are already available. *Then,* create a communication plan aligned with the resources you have.
- Determine the right tool, media, venue, and frequency for each communication.
- Practice how you share knowledge verbally. Are you expressing Knowledge Statements and citing your sources? Are you providing market context?

Chapter 14
PREPARE FOR CHANGE

Prepare for Change

In 1915, Alabama experienced a boll weevil infestation. Boll weevils are beetles that infest and feed on cotton crops and, as cotton was a major cash crop of the area, the local economy was decimated. Farms and families went bankrupt, and extreme poverty caused a variety of social challenges in the area, all in the space of one growing season.

Two citizens of Coffee County, Alabama, saw this crisis as an opportunity to convert the area from cotton to peanut farming and other agricultural best practices pioneered by famed Tuskegee University researcher George Washington Carver.

The first peanut crop was successful, and more farmers planted peanuts. Cotton was grown again, but farmers learned to diversify their crops, a practice which brought new money to Coffee County.

The town erected a monument to the boll weevil as a tribute to how something disastrous can be a catalyst for change, and a reminder of how the people of Alabama adjusted in the face of adversity.

Change can be a nasty thing, and changes can disrupt far more than our businesses. We could get really depressed! Instead, let's prepare to overcome them and win.

Change can happen slowly or incredibly fast; it can impact your business, your market, or the whole world. How you respond to change shouldn't be an accident; maybe it's even the opportunity to blaze trails, improve your position in the market, or open an entirely new market. Meteors are often tragic social, cultural, or well-being events such as COVID-19 or September 11, 2001. Even when they only apply to you and your organization, the impact is massive. As such, they require a particular workflow and prioritization unique from "standard" listening. This concept is significant enough that I've given it its own chapter later in the book.

I see innovation and market change as the yin and yang of most strategic business activity. In this chapter, let's explore how you can actually plan for change—even when you don't know what change is coming—and how you can innovate on purpose when it comes.

A DEFINITION OF "CHANGE": THE METEOR

First, let's get clear on the type of change I'm talking about in this chapter. What I'm not talking about are the changes you should be catching in your day-to-day market research activities. New regulations, a market that's beginning to shrink, or the latest systems your product should interoperate with are all market changes you should be aware of if you're using the workflow we've discussed so far. The changes I'm talking about are unexpected, severe, and requiring an out-of-the-norm review and response.

Nassim Nicholas Taleb popularized the term "Black Swan" in his five-volume series *Incerto*, of which the best-known books are *The Black Swan* and *Antifragile*. *The Black Swan*, about the nature and impact of unpredictable events, was published in 2007 as the second of the series. The book has been credited with predicting the banking and economic crisis of 2008. Taleb cites three characteristics of a Black Swan:

- A very rare, unpredictable event—an outlier in our common history.
- An event with an extreme impact on the world or a large part of it.
- An event that seems like it *should* have been predictable but was not predicted.

The change we're talking about isn't necessarily a Black Swan as defined by Taleb, but it *is* a significant event demanding review and response. I prefer to think of them as **Meteors**.

Merriam Webster defines "meteors" as bodies of matter from outer space that enter the earth's atmosphere, becoming incandescent as a result of friction and appearing as a streak of light. The difference between a meteor and a meteorite is that a meteor becomes a meteorite when it hits earth. Our goal? To resolve the meteor before it becomes a meteorite.

Our Meteors include events that are a little less extreme than Taleb's Black Swans, although we will include those of course. Here are the criteria we'll use to define Meteors:

- *Unusual.* The change is outside the realm of regular expectations. It's not something you see every day. Whether it's a financial downturn, the rise of a new technology, or a boll weevil infestation—they're singular events. They don't necessarily have to be societal or large-scale; it could be a company disaster. The Tylenol bottles that got tampered with in 1982 pretty much only impacted Tylenol, but it was certainly an outlier for them.

- *Unpredictable.* When we look back on these changes, we feel like we should've seen it coming, but the fact is, we don't. While I'm going to give you some tools to reduce the number of surprises and mitigate their severity, these are events in your world that you don't expect to happen.

- *Unplannable.* Because each specific meteor is unpredictable, you can't have a plan ready for every possible meteor that could come at you. We'll discuss how you *do* get ready for meteors.

- *Impactful.* I'm going to say that if something unexpected threatens *your* viability as a business, you can think of it as a Meteor, even if it's not impacting others.

An example here is what I call the Google Effect. Google or some other massive company decides to offer your product for "free," or bundled or embedded in something else. That may just affect you, or perhaps your industry, but not be widely known.

- *Requires response.* These Meteors need to be resolved before they impact your business. The action you take may be to wait and see for a few days or stop selling your product right now. Something needs to be done, however, and because the Meteor is a big deal by definition, the pressure to get it right is huge.

- *Not always bad.* The good news is, like Black Swans, a meteor could offer amazing opportunities for your organization. Upheaval almost always brings risk or bad news to some and opportunities or good news to others. During the COVID-19 lockdowns of 2020, brick-and-mortar shops were devastated, but online shopping and delivery services thrived. Netflix, the iPod, and many other products were successful at least in part because of technology Meteors that created an opportunity for innovation.

- *Self-defined.* Because this book is intended to lay out a workflow for *you* to deliver exceptional innovation, we're focused on you and *your* challenges. As I mentioned earlier, Tylenol certainly thought the tampering issue of '82 was a Meteor, even if Bufferin or NyQuil didn't. If you think it's a Meteor, it's a Meteor. The question is, how do you create a workable definition so you can plan? We'll come to that in a bit.

Meteors are market or ecosystem events that you can't plan for but must react to. Meteors can be discovered through your innovation workflow efforts, but even those that are complete surprises require a workflow and team to assess and address them.

From COVID-19 to natural disasters to something that happened only to you and your company, I'm sure you can think of a few events or issues that you'd consider Meteors. Now that you know what they are, let's see what we can do about them.

STEPS TO CRUSH THE METEORS IN YOUR WORLD

I've spoken at a couple of conferences on this topic, and the question I get most often is, "If a Meteor is defined by being unpredictable, how can I possibly *plan* for it?" Not to mention it seems like a huge waste of time to sit around and worry about things that may never happen, when we have plenty of problems right in front of us!

But planning for Meteors may be the work that saves your business, or even enables you to become the singular option for your market for the simple reason that you're ready to serve them and your competitors aren't. In other words, planning for Meteors is the last piece of a workflow to innovate on purpose. And, like the other pieces, it shouldn't be time consuming or onerous. What we're going to plan for isn't the Meteor, it's how you'll address it once you see it coming.

I'm going to give you five practical, resource-effective steps to be ready for *any* Meteor that heads your way, beginning with preparing before the challenge is even on your radar:

#1: INVEST IN MARKET KNOWLEDGE *BEFORE* A METEOR APPEARS

Whether or not you're a prepper in your personal life, it pays to squirrel away knowledge of your market and your place in it *before* something crazy happens. Every organization does some market research, and I hope by now yours is intentional, efficient, and based on who you're trying to become for those markets. Remember this: The more you know about your markets before a Meteor hits, the better positioned you are to react afterward.

Here's a great example of how two different airlines reacted to a recent Meteor:

The year was 2020, and the Meteor was COVID. Airlines had to deal with near-total stoppage of travel—specifically business travel.

And they were faced with a unique challenge: how to keep their best customers—the loyalty-program road warriors—loyal?

For frequent flyers heavily committed to one or two loyalty programs, COVID's disruption of in-person meetings requiring air travel created a unique concern. The thinking was: I fly all the time for my job and assume I will again. Will I have to spend a year rebuilding my frequent-flyer status? Of course, there were bigger problems for everyone to worry about, but "losing status" was a concern in the context of the airlines and the frequent flyers.

Being a full-time instructor with Pragmatic Institute at the time, I was used to flying out-and-home nearly every week of the year. I had *major* status on two airlines. In April of 2020, within a few days of each other, both airlines sent me a marketing email.

Here's a paraphrase of Airline #1's message: "The last thing you need to be worried about right now is your frequent-flier status. So don't. We're extending your [super-deluxe status] through 2021. We look forward to when we meet again."

The message from #2: "Earn double miles on trips from April through June!"

How could one airline be so tone-deaf to its loyalty-program customers and another so completely in tune? Because one *already knew* its market better than the other.

The best time to start blasting Meteors is before they show up. The investments you've made so far in innovating on purpose will stand you in good stead to react well—whatever crazy thing falls from the sky.

There are a few *specific* questions you should get answered before a Meteor shows up on your horizon. If you don't already have Recognized Market Knowledge Statements on the following questions, add them to your backlog:

- *What alternatives do your markets have to your product?* This is a question you should already know the answer to in order to position your product effectively in the market. But in the event a Meteor hits, understanding what your market has to fall back on if you're out of commission prepares you for action, whatever the Meteor may be.

- *Why do they buy your product and not someone else's?* I searched for "beauty subscription boxes for women" on Amazon and got 152 offerings: for redheads, vegans, those wanting to be beautiful through cruelty-free products, over 50 beauty and date night glam. Those are obvious market differentiators for the day-to-day decisions a buyer might make. But in addition to the obvious, do you have something that makes you unique to your customers that could keep them buying from you even if a vegan-product competitor comes along that's cheaper than you? Additional advantages may make a difference in unusual market situations.

- *What would make them stop buying from anyone?* It's pointless to try to anticipate a specific type of Meteor, but it's powerful to know what would have to happen to make your buyers stop buying entirely. In the B2B software world, it's possible that your market would be able to get along without your product, so their threshold to cancel or not buy at all could be a matter of a slight economic downturn. In others, you may be so critical to their operations that as long as they exist, you exist. Getting a sense for how vital you are to the markets you serve gives you an order-of-magnitude comparison to whatever inbound Meteor's heading your way.

- *Where are the strengths and risks in your supply chain?* Your supply chain could be your biggest risk if a Meteor hits (no news there, right?). Hard goods, services, delivery, fulfillment, cloud providers—how necessary and risky are each?

- *Would different objects answer these questions differently?* Would adult children and their parents, executives, and individual contributors, lawyers, and professors have different priorities and needs? Would they have different options even if the same Meteor hit all of them? Priorities and power in the decision are very different depending on the product you offer, the alternative available, and the Meteor itself. The more complex your business and the more disruptive the innovation phase you aspire to, the more objects you'll have to learn about to prepare.

Whoa—that's a lot of questions! But if you look at them carefully, you can see the benefit in having this knowledge right now, in your business-as-usual workflow. You can also see that getting the answers to such questions might be as easy as doing a single survey of each object you need answers for.

And you'll need this knowledge in the event of a Meteor—any Meteor—because when you see it coming, you'll have to leverage whatever current knowledge you have about your market in your initial response. Solid market knowledge and you're Airline #1. No idea? Airline #2.

Let's move on to #2.

#2: ADDRESS THE OUTCOME, NOT THE METEOR

METEOR

Meteors

GREAT LAKES FLOOD
MICHIGAN

This is really a shift of perspective. When a Meteor appears, it's natural to focus on the Meteor itself and to think about how to respond to that unique event. But since Meteors are events you can't imagine ahead of time, you can't prepare for them specifically. You certainly can't plan for every type of Meteor you could think up either! But there *is* a resource-effective way to react quickly to the unexpected. You'll respond more effectively to a Meteor if you stop thinking about the Meteor itself and look to the **outcomes** that Meteor could drive—and the opportunities it creates. Let's explore that a bit using an example.

A Meteor is the event that *causes* all the problems: COVID-19, the rise of cloud computing, a chip shortage. Let's assume, as an example, that the whole state of Michigan has flooded. Who could guess that would happen? What could we possibly do about it now that it has? Meteors are unique—literally unimaginable before they occur and likely impacting you in a variety of ways. Let's say you're a "special occasion" gourmet food subscription box business: If you try to "solve" the Meteor, chances are you'll spend a lot of cycles to very little effect.

METEOR   OUTCOME

GREAT LAKES FLOOD   LOCAL GROCERY
MICHIGAN   STORES ARE CLOSED

Meteors and Outcomes

A better way of thinking is to observe what *outcomes* the Meteor is causing. In our Michigan example, an outcome of the flood is that grocery stores are closed. People who are struggling to save their homes now have to worry about how they're going to eat. And although they still have to eat, they may not be whipping up fancy meals during the rebuild, which means they may suspend or even cancel their subscriptions.

You *can* respond to outcomes—they may even provide big opportunities for you to help your current customers out of a dire circumstance! Maybe you could deliver ready-to-eat meals and other staples instead of gourmet meal ingredients—by boat, of course.

METEORS, OUTCOMES AND OPPORTUNITIES

METEOR	OUTCOME	OPPORTUNITY
GREAT LAKES FLOOD MICHIGAN	LOCAL GROCERY STORES ARE CLOSED	"BASICS" FOOD BOX SUBSCRIPTION DELIVERED BY BOAT

Don't worry about the Meteor. Instead, brainstorm outcomes you could face and determine your current readiness for those. To keep this pertinent to you, brainstorm the outcomes your buyers said would result in them no longer buying from you. The good news? Even if you don't have the resources to brainstorm what outcomes *could* occur, this change in perspective gives you the tools to react to them more efficiently when they *do* occur.

Examples of Outcomes to Prepare For

The very nature of a meteor is that it's something unusual and unpredictable, so there is no complete and definitive list you can plan for. That said, these are some of the types of meteors—and outcomes—you should probably expect and plan for:

The Market No Longer Needs What You Offer

Oh, come on—wouldn't we see that coming from months or years away? I would argue that many businesses put their collective head in the sand as their products slowly obsolesce, but that's a discussion for a different book. In this case I'm talking about a rapid, massive change in priorities or needs—the outcome of a Meteor.

An example from my own experience had to do with my time in the publishing world. Several years ago, I led a product and marketing team that dealt in content rights brokerage. Essentially, our organization represented the re-use and republication rights to a large body of scientific, technical, and medical research articles from the top industry journals. We brokered those rights to, among others, universities that wanted to distribute the materials to students without running afoul of copyright laws, a practice called, "fair use."

That said, those laws were open to interpretation, and universities often felt that the interpretation publishers (and our organization) put forward were restrictive, stopping them from distributing knowledge their students needed because of the publishers' desire for monetary gain.

In 2008, a lawsuit was filed by a few publishers against a particular university and was decided in favor of the university in 2012. (Since that time the ruling has been reversed, and many other legal and cultural changes have changed the landscape.)

A talented product manager I'll call Jack was responsible for the products sold to universities. Jack spent a great deal of his market research efforts focused on this lawsuit, even though our organization was not a plaintiff or defendant.

Why? How much could the lawsuit of a small group of publishers against one university impact us? Plenty. Jack was monitoring chat groups and reaching out one-on-one to our current university customers, and found that most universities in the US were watching this lawsuit with great interest. He summarized the comments of one university representative he was speaking with as: "If this suit is settled in favor of [the university], we won't be buying Product X from you anymore." If they no longer needed permission to use the content, they no longer needed our product. Yep—that officially qualifies as a Meteor!

But because Jack was watching the suit unfold and keeping track of the buzz, he knew a big chunk of his revenue was at stake and kept me and the rest of the executive team apprised of the situation. At one point, we reprioritized some product development to get one of his backlogged initiatives out in the market faster, in order to mitigate the anticipated revenue loss/backlash from the lawsuit. As much as we would've liked the new product to deliver incremental gains, it was enough to compensate for the loss in revenue that did indeed occur as a result of the initial court ruling.

Had this ruling been a surprise to Jack and the organization, the consequences may have been dire—not just in terms of revenue but of lost industry confidence and certainly loss of confidence in me and my team. In this case, the revenue couldn't be salvaged, but the losses incurred could be prepared for and communicated. Thanks to Jack's focus on his market, the executives and the board were able to plan and pivot, not without loss but without panic.

Market Alternatives Change

People's taste change over time as a natural part of our world. This usually happens slowly and due to a wide variety of factors, but sometimes it happens fast, as a result of a Meteor. Here's an example of both: golf as a pastime in the United States.

In 2019, there were more golfers over the age of 70 than there are under the age of 40—a big departure from 20 years before. While not a dying sport, it was certainly in a decline. There were many reasons for this: new, inexpensive alternatives such as disc golf; fewer informal business meetings happening on the links—a variety of cultural shifts that were shifting us away from taking up the sport.

What seemed like its death blow came in early 2020, when Americans in many states were prohibited even from playing non-contact, outdoor sports together. Golf courses remained empty all spring. But by June 2020, also because of the COVID Meteor, Golf's *Datatech* reported a 14% **increase** in rounds played in June year-over-year. The total number of rounds played in 2021 bettered the industry's previous high set two decades ago, despite fewer golfers and fewer golf facilities. Whether the increase will hold in the coming years is something *all* leisure activity providers should be keeping an eye on.

Conversely, Netflix received 15.8 million new subscribers during the first quarter of 2020, a massive increase compared to the expected

7 million. In 2021, Netflix lost subscribers for the first time in its 14-year history.

Yes, this happened because of the Meteor of COVID, but it was the *outcome* of having more, or fewer, options for what to do in their spare time that created these swift market changes.

How quickly could you ramp up to take advantage of being the only option for your market? How will you know if the options are about to shift? Right after 9/11, there was a run on American flags. Stores were out of stock instantly, with none to be had for weeks. In December 2022, Southwest Airlines had a major software meltdown that resulted in them cancelling nearly all their flights.

Someone Starts Offering Your Product for "Free"

Who doesn't love Google Maps? But before I had that magical app on my phone, I carried around a purpose-built device that helped me find my way around: a Garmin. Of course, you had to pay for updates to the Garmin software every year if you wanted to find half the roads in Austin or any other high-growth city, not to mention paying for the device in the first place. Why bother when there was a "free" option I could download onto a device I carry anyway?

How can you compete with "free"? The case of Garmin is one of the stronger "pivot" case studies in decades, and I recommend reading it in detail. But the main point here is, everyone is at risk of what I call, "getting FAANG'd": the outcome of one of the Big Five tech companies (Facebook, Amazon, Apple, Netflix, or Google) deciding to flip your revenue, distribution, or buying model on its head and shut you down.

At this writing, Google has just launched an education/ certification program pointed at tech skills and has already received commitments from several organizations (including, of course, themselves) to hire graduates of these programs. How will this disrupt education? We don't know yet, but we do know that they've synthesized a few broad-ranging market and cultural trends and responded to them.

Other Examples of Outcomes

- ***The market is shrinking.*** Several years ago, I worked for an organization whose primary market was print publishers. At the time, that market was shrinking rapidly, and my company—and industry as a whole—was in a panic. That said, it hadn't occurred to them to look to other markets to shore up revenue. Rapid decrease in market size may be an outcome to brainstorm. A decreasing or unstable market could be the outcome of slow societal shifts or the rapid trajectory of a Meteor; in either case, consider expanding into newer and stronger markets.

- ***The market is getting harder to serve.*** Perhaps your market is becoming much more heavily regulated than in the past. Perhaps M&A activity has rendered an industry you've served for decades chaotic and unprofitable. Perhaps the market you serve is on the decline, with budgets for anything but the most necessary tools being cut. Whatever the reason, if a market gets hard to serve, you might want to look around options that aren't.

- ***Market demographics stay the same but habits change.*** What if you keep thinking of your market as behaving certain ways just because they are a certain age? It's what I call "legacy market knowledge" and can put you at serious risk of disruption from a competitor that's more in touch with the market. Just because you've been selling to 10-, 40- or 70-year-olds forever doesn't mean you know what people that age want now, unless you actually listen to them.

You can't anticipate every type of Meteor you may encounter, much less prepare for them, but you *can* respond to outcomes—they may even provide big opportunities for you to help your current customers out of a dire circumstance! Don't worry about the Meteor. Instead, learn from your market the outcomes that could put you at risk and brainstorm how you could go after the opportunities they provide. And be sure to telegraph into the organization that something might be changing by changing any "Recognized" market knowledge to "Transitioning" status.

The good news is, if you've used the tools I propose throughout this book, you're already looking for Meteors; you just didn't know it. What if you find one?

#3: MAINTAIN A MONITORING AND ESCALATION PLAN

Imagine—tomorrow you wake up and there's no Internet. None. No Zoom. No Slack. No Outlook. What do you do? How do you communicate with your team? With your customers?

Meteors are outliers, every one will be different, but putting a plan in place to monitor the horizon and act quickly in the event one is blazing your way is a good decision. This is the section where we tackle the "act quickly" part of the equation.

What Is an Escalation Plan?

The day-to-day execution of your contributing strategy, using your Innovation Map to guide you, should turn up more than a few Meteors without having to do any extra work at all. But, once discovered, how do you decide what to do about them?

Because Meteors, by definition, demand a response (usually a fast response), this is no time for a backlog or roadmap. You don't have time for them to go through your usual workflow. I recommend putting Meteor Team in place, with an escalation workflow used by the whole company. Like the volunteer fire marshals in an office building, this team and workflow will very seldom be called upon. But when they are, they're ready.

Escalation: Team Responsibilities

While the actual Meteor Team will be small and nimble, every pair of eyes in the company should be trained to recognize a Meteor and know what to do if they see one.

- *Everyone is responsible for sighting Meteors.* In the organization should be able to report Meteors to the Meteor Team. This doesn't have to be complicated, but make sure you have more than one reporting channel in case an outcome of the Meteor is, actually, the Internet being down.

- *Meteor Team acts as a stage-gate for review.* All potential Meteors are funneled to the Meteor Team. My recommendations for assembling your Meteor Team include:

  Choose no more than 10 senior, cross-functional representatives. These folks should be chosen by the company owners/execs but not include them. This is a stage-gate team. If you're not a senior-level person or your organization is so large as to make this unwieldly, choose your own team for your specific areas of responsibility.

  Those chosen for the team should be creative thinkers, engaged employees, and precise communicators. Ideally, you bring as many skill sets to the table as possible with a small group so that any potential Meteor reported can be assessed from many perspectives.

  The role of the Meteor Team is to review the potential Meteors submitted by members of the organization, to

An Escalation Workflow Enables You to Act Fast When Something Changes

AN ESCALATION WORKFLOW ENABLES YOU TO ACT FAST WHEN SOMETHING CHANGES

Formal and informal monitoring delivers potential Meteor sightings

Potentials are delivered to the Meteor Team, who brainstorm recommendations

The Review Team approves, mandates resources, communicates with owners

determine whether it is a Meteor, what the outcomes could be, and what options are there to either mitigate the risks or capitalize on the opportunities. The Meteor Team should take recommendations to the Review Team.

- **The Review Team clears the way to action.** The Review Team acts as a check on the Meteor Team and, unlike the Meteor Team, the Review Team should be *very* senior—able to speak for the company and owners, able to immediately make resourcing and prioritization decisions and mobilize anyone in the organization for help. A logical Review Team would be CEO, chief legal counsel, head of IT, and head of product. If action is deemed necessary, the Review Team appoints the Solve Team.

- **The Solve Team gets it done.** If approved, action is quick but deliberate and specific to the Meteor, outcomes, product, market, and severity of the challenge. Whoever is taking the action is basing it on market and business outcomes, what's known right now, and what you need to find out. This is a great time to run the type of hypotheses and experiments we discussed earlier in the book.

    Quick cycles of experimentation in the market, paired with the right type of market feedback loop, ensures that you can iterate as the situation evolves and move in the direction your market is headed.

#4: HYPOTHESIZE AND VALIDATE IN THE MARKET

It's only logical to take a stripped-down, speeded-up version of your usual workflow to the market when a Meteor is on the horizon. But a question I get asked a lot is, "How am I supposed to get customers who are afraid for their lives to talk to me or participate in an experiment—how can I get information if the Meteor is coming at all of us right now?" Here are a couple of ideas based on my personal experience and discussion with clients and students:

- *Give them a deal. A good one.* Because Meteors hit fast, are unexpected, and require a response of some sort, now is not the time to be stingy. You need to learn and you need to learn fast. Your market may not see that as a priority; if there's any way that a good deal on a new product module, or a Beta version of something will get them to try it, ask your Review Team for the funds and approvals to get it done.

- *Test recovery options before you roll them out to everyone.* I'll bet Airline #2 in the example I cited earlier wishes they'd done that.

- *Update your cancellation reason codes* and other standard feedback loops to gather what's meaningful right now.

- *Get expert insight* from applicable groups in authority, whether it's FEMA, industry associations, parent groups or others, and monitor updates as they come through.

- *Offer to share the results of surveys* or other research with the sources you reach out to; they will appreciate it.

- *Finally and perhaps best: lend a helping hand!* A big outcome from COVID was the depression and sense of unreality felt by the 30 million Americans who lost their jobs practically overnight.

Headspace, a meditation and mindfulness subscription, offered free subscriptions to US workers who were unemployed or furloughed. The free subscription included a complete library of meditation and mindfulness content for that year.

Headspace didn't have to do this, and their main goal was not to pick the brains of those who signed up. But it was admirable and authentic. And by bringing them in with a free offering, they could learn while doing something good.

Genuine offers to help in a crisis are remembered long after the Meteor is smashed, but relationships you build during crisis time helps you better understand how your market is changing right now and establishes the goodwill to continue the dialogue into the future.

Learning is great, but not if you don't use it. So let's move on to the fourth recommendation for dealing with change.

#5: DO A RETROSPECTIVE

You've crushed the Meteor! Now what? It's time to take a breath and learn.

How many of you did a retrospective on your response to the last Meteor you faced, whether it was COVID or the bird flu? Did you hold a few meetings internally and with the market to document what went well and what could be improved next time? For those of you who did, how many of you are *using* the knowledge?

There's a classic tool used by military and business the world over to articulate lessons learned after any action taken by a team: the After-Action Review. It's very useful to learn from your response to a Meteor.

> The US Army has been using After-Action Reviews, or AARs, for decades. AARs compare the **intended** execution and outcomes of an action to the **actual**. AARs are forward looking, with the goal of informing and improving future planning, preparation, and execution of similar actions. An AAR is tightly focused on participants' own actions; learning from the review is taken forward by the participants. Recommendations for others are not produced. Finally, assigning blame or issuing reprimands is *not* the purpose of an AAR. The purpose is to learn so everyone can do better next time.

On September 26, 2022, NASA altered the orbit of the meteor Dimorphos with the DART spacecraft. This was a test to see whether the spacecraft, guidance systems, and theory itself would be able to alter the orbit of a future meteor that was heading toward Earth. Yeah, okay—I know that was an asteroid and not a meteor, but it's still a good analogy. The lesson? Create the tools and workflow that will help you address whatever falls out of the sky at you–before you can see it coming.

ACTION PLAN FOR CHAPTER 14

- List the Meteors you've faced in the past 12 months. How could you have handled them differently? What Meteors are you facing today?

- Refer to the list above: What were the outcomes of those Meteors that you could react to, versus the nature of the Meteor itself? How does this way of thinking help you act more effectively?

- Identify your escalation team and create an escalation plan.

Chapter 15
FINAL THOUGHTS

You and your company are already awesome, and you already have a lot of what you need to take that awesomeness to the next level. Don't let what you already know about how to crush competition and lead the way to the future get lost because you're not managing it. Don't let the blind spots in your knowledge, or lack of alignment with leadership, create roadblocks you don't even realize are stopping you.

There are plenty of roadblocks to successful innovation that we can't control. You may not even be able to impact those self-inflicted by your *own organization*. But you can smash through the roadblocks you yourself set up. As a final thought, I recommend removing these temptations from your own responsibilities:

DON'T CONFUSE RECOMMENDING WITH DECIDING
I've often said—and thought—that I was the CEO of my products. I was wrong. The CEO is the CEO of the company and therefore the CEO of the product. We are, however, in a role of massive responsibility and power, if we choose to use it wisely.

The role of the Market Strategist is to be the source of truth on the market, within that Strategist's sphere of responsibility. That's an awesome responsibility, but it's not an absolute one. Your CEO may choose to take your market-driven advice, or they may choose to go another way. Remember that your role is to bring the best market information to the table to justify recommendations, learn as much about any decision that's made by leadership, and enable others to execute.

DON'T MANIPULATE MARKET DATA
TO JUSTIFY WHAT YOU WANT TO DO

As the central source of knowledge (you see I avoid the use of the words "truth" and "facts") on the markets your organization serves, you have a sacred trust. You are trusted to obtain and present the information from your market to the very best of your ability—the whole knowledge, and nothing but the knowledge. If you ever catch yourself looking at new insights and saying, "Yeah, I guess they hate our 50% up-front fee, but they don't know how often we get stiffed!" and adding that editorial, or sliding that insight to the back of your mind, you're doing a grave disservice to your organization as well as your markets.

Passionate dispassion is what I call it. As the Market Strategist, you can't get attached to what is, or what you think, or what your organization did to become successful yesterday. Markets change— their needs change, their priorities change, and how they decide how to solve their problems changes. If you hold onto ideas that the market has moved past, the market will move past you.

Be passionate about serving a market, but don't fall in love with your products. You are passionate about building great solutions but not about any *particular* solution. You are passionate for your products to succeed, but understand that as market problems and priorities change, so must the solutions you build to solve them.

You're the market expert, but not because you personally have been in the market or have the most experience with your company's products but because you're out in the market gathering firsthand knowledge of their needs continuously, as well as bringing together all the sources of knowledge needed for a complete picture of the market.

DON'T MEDDLE IN THE WORK OF OTHERS

Truly, the reason most organizations don't have enough market insight is because they expect and encourage their Market Strategists to do the jobs of others. Embed yourself in the development team—help them out! And we love to put together design suggestions and give tips on how to create sell sheets and marketing campaigns. But (and you all know this) the more time you spend with the execution experts, the less you empower them to innovate on purpose.

It's not you they need to hear from—it's the market, **through** you. Further, the Market Strategist is not the person who translates market needs into fulfilled products but rather the person who is finding the "what's next?" Without you gathering, aggregating, analyzing, and prioritizing the needs of the market, development will build what seems cool to them, marcomm will create beautiful, but perhaps tone-deaf, materials, and sales will focus on renewals because nobody wants the new products.

DON'T STOP HAVING FUN

The role of Market Strategist is a risky job. It's a stressful job. But being a Market Strategist is the best job any organization offers. Enjoy your work and the results you achieve.

It's time to take the practical steps laid out in these chapters and innovate on purpose at the intersection of what your market needs and who you want to be. Define a scope, do a scavenger hunt, and build opportunity statements that guide you to your next step in market and product success.

Good luck!

About the Author
DIANE PIERSON

As a product and marketing executive, Pierson delivered over $100 million in new revenue for B2B software, data and analytics companies around the world. Her work has been recognized with numerous CODiE awards and more than one appearance in *KM World's* List of Trend-Setting Products. Most recently, Diane was a full-time instructor, coach and thought leader with Pragmatic Institute.

Today, Pierson is an authority on strategic product and marketing innovation at the intersection of what your market needs and who you want to be. She's the founder and Chief Market Strategist at Innovate on Purpose, a consultancy enabling market success by helping organizations focus their product and marketing efforts on the right ideas with a proven, five-step workflow.

Pierson has a BA in English from Kalamazoo College, an MBA from DeSales University, and executive education from Babson College and The Wharton School. She and her husband Tom live in Austin where they're always on the lookout for the best breakfast taco in town.

Innovate on Purpose is Pierson's first book.

www.ingramcontent.com/pod-product-compliance
Lightning Source LLC
Chambersburg PA
CBHW060548200326
41521CB00007B/531